ELECTRONICS IN HAZARDOUS AREAS

Ian Staff

ELECTRICAL TRAINING CONSULTANT

First Edition published 2023

2QT Publishing Services

Stockport UK

Cover design: Dale Rennard

Images supplied by author

Printed in the UK by Ingrams UK

ISBN 978-1-914083-91-4

About the Author

I am an Electrical Training Consultant and carry out Electrical Training for a company in Hull by the name of Humberside Offshore Training Association Ltd. (H.O.T.A.), Malmo Road, who, by the way, is one of my sponsors. Before my fifteen or so years at H.O.T.A. as a Trainer /Assessor I spent thirty-eight years with BP; seven of those as their Instrument/Electrical Training Officer in charge of all Instrument and Electrical Training and part of their Team in their Training Department where I obtained my Training and Assessing Qualifications.

The Maintenance Electrical Technician in Hazardous Areas (Full Set)

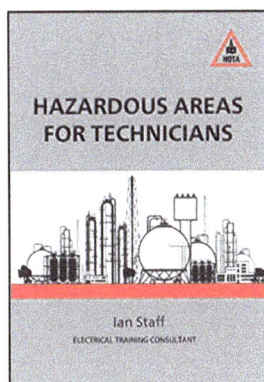

ISBN 978-1912014958

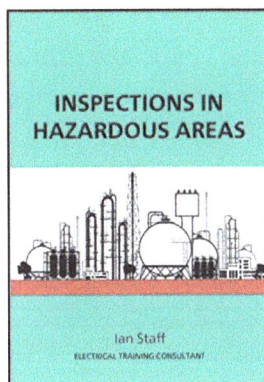

ISBN 978-1913071615

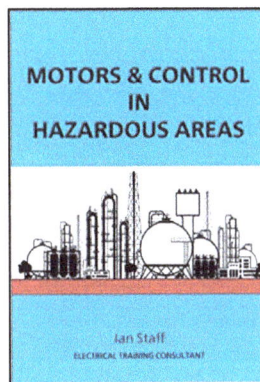

ISBN 978-1914083013

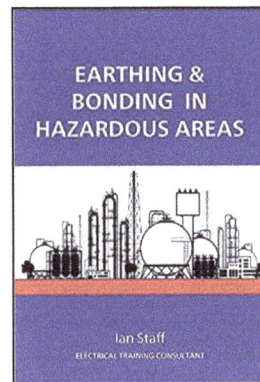

ISBN 978-1914083112

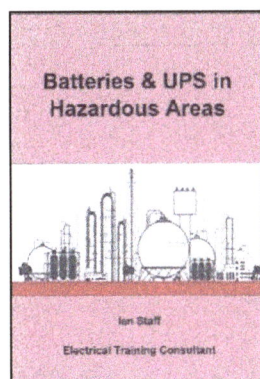

ISBN 978-1914083303

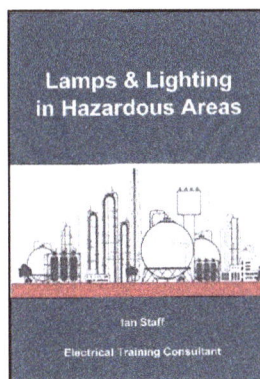

ISBN 978-1914083662

So far I have also written the above six books in this set showing the knowledge that, in my opinion, is required by an Electrical Technician on a Chemical Factory or a Platform. Having once been there I feel that these books, at a reasonable cost, written at Technician level would demonstrate the full spread of subjects dealt with in everyday life.

The first book talks mainly about Atex, the IEC and EN/IEC Standards, Zones, Gas Groups, Temperature Classes etc. and sets the basis for the other five books; Inspections of Electrical and Instrument Equipment, Motors and Control Circuits, Plant Earthing and Bonding System Types and Testing, Battery types and Maintenance including UPS Systems and finally Lamps and Lighting Systems shows all of the different Lamps/Light Bulbs available, where they are used and controlled in hazardous areas.

Introduction

This book: 'Electronics in Hazardous Areas' is the seventh book in the series and explains the history of Electronics as well as diagrams and explanations of all Electronic Components that are available.

My first book 'Hazardous Areas for Technicians' is very successful, explaining what Atex is and how Atex equipment is installed and maintained in Hazardous Areas as well as things like 'Cathodic Protection' etc.

My second book 'Inspections in Hazardous Areas' is mainly for Companies who carry out their own Inspections of Hazardous Area Electrical, Instrument and Mechanical Equipment and focuses how Electrical, Instrument and Mechanical Inspectors go about understanding standards, why the equipment is inspected and the terminology.

My third book Motors and Control in Hazardous Areas is all about the History of the electric motor and all of the different motors that Electrical Technicians may come across in their Hazardous Areas from DC to AC.

My fourth book 'Earthing and Bonding in Hazardous Areas' which like the others starts off with the history of the colours and symbols right up to how to carry out the installation and testing and why various tests are done. The book explains the importance of the Earthing and all of the different methods that can be installed into the ground to achieve this.

My fifth book 'Batteries and UPS in Hazardous Areas' deals will all of the different batteries that are available, how they work and where they are used. UPS systems that companies may have installed are also discussed.

My sixth book is all about lamps and lighting systems, how the different lamps work aided with diagrams. Lighting systems are explained and the dangers involved in fault finding.

Content

Electronic History:

Different Electronic components were developed by various engineers and physicists all the way through history and we must be thankful to them for what we have in Electronics today.

Let us have a look at a few characters of history and the contributions they claim to have made; you may know more.

When you are looking through just remember what year it was when they developed their equipment.

1745 – Ewald Georg Von Kleist – German Scientist invented a device for storing charge.

1745 – Pieter Van Musschenbroek – Professor of Maths Invented a device called a 'Leyden Jar' which was the predecessor of the Capacitor.

1795 – James Clerk Maxwell – Discovered the theory of the Diode.

1827 – Georg Simon Ohm – German Electrician discovered Resistance.

1834 – Charles Babbage – Invents a mechanical computer. (Analytical Engine.)

1873 – Frederick Guthrie – British Physicist. Conducted experiments and explained the principle of a Diode.

1904 – John Ambrose Fleming – Developed the Diode on a commercial scale.

1906 – Lee de Forest – Invented the Triode.

1909 – William Dubilier – Invented the Mica Capacitor.

1912 – Edwin Howard Armstrong – Developed the Oscillator.

1935 – Paul Eisler – Austrian Engineer invented the Printed Circuit Board.

1947 – J. Bardeen & W. Brattain – Scientists at AT & T's Bell Laboratories are acclaimed to have invented the first Transistor.

1948 – W. Shockley – Scientist at AT & T's Bell Laboratory claimed to have invented the first Bi-Polar Transistor.

1950 – Clarence Melvin Zener – Professor – developed the Zener Diode.

1951 – The first commercial computer was invented.

1956 – IBM's RAM (Random Access Memory) was invented.

1957 – Thyristor – Introduced by GEC.

1958 – Jack St. Claire Kilby – Invented the first basic Integrated Circuit.

1962 – Hofstein & Heiman – Invented the MOSFET Transistor.

1959 – Jack St. Claire Kilby – Received a patent for an Oscillator.

1959 – Robert Norton Noyce & Jack Kirby – Developed another Semiconductor Chip.

1964 – Douglas Engelbart – Invented the computer mouse.

1968 – Douglas Engelbart – demonstrated a prototype computer with a mouse.

1971 – Ted Hoff – Developed the first Microprocessor.

Atomic Structures:

Before we can fully understand the workings of our electronics we need to know a little about the atomic structure of materials so that the insulation, conduction and semiconduction of materials used, and why inventors used them in the first place, can be better understood. I hope that I have not made the following explanations too complex.

Molecules:

A '**Molecule**' is a small unit of an '**Element**' or '**Compound**'. Simply we can say that a molecule is a group of two or more '**Atoms**'. The structure of a Molecule is held together by what is called '**Covalent Bonding**'. So for instance if we take water which has a chemical formula of H2O, what this is telling us that it is a molecule made up of 3 atoms; two of Hydrogen (H2) and one of Oxygen (O). Carbon Dioxide CO2 is one Carbon Atom (C) bonded to two Oxygen Atoms (O2).

Another formula might be **2H2SO4** (two Molecules of Sulphuric Acid) so what would be the meaning here? Well the first large 2 means that there are two Molecules (We know that Molecules contain two or more Atoms) and the next figures and smaller numbers show there are seven Atoms. In this case, as mentioned, the chemical symbol is for two Molecules of Sulphuric Acid (2H2SO4) containing two Atoms of Hydrogen (H2), one Atom of Sulphur (S) and four Atoms of Oxygen (O4).

Atoms:

So where does the above leave us with Atoms? We can say as a definition that an Atom is the smallest unit of an element or compound. Atoms are composed of a nucleus which has Protons and Neutrons inside and Electrons in orbit around the nucleus. The number of Electrons will always be equal to the number of Protons and this will be their '**Atomic Number**' on the **Periodic Table of Elements** with some examples given below.

1	19	29	13
H	**K**	**Cu**	**Al**
Hydrogen	Potassium	Copper	Aluminium
Non Metal	Alkali Metal	Transition Metal	Post Transition

The letters on the Periodic Table, left, do not always seem to denote the Element. It tends to pick out the older names for Elements i.e. 'K' for Potassium comes from Kalium, which is Latin for Potash.

Looking at the diagram above of sections from the Periodic Table we see Hydrogen (H) has one Proton and one Electron, it is a **non-metal**. Potassium (K for Kalium) has nineteen Protons and nineteen Electrons and is an '**Alkali**' Metal. Copper (Cu for Cuprum) has twenty-nine Protons and twenty-nine Electrons and is a '**Transition**' Metal. Transition meaning that they are in the centre (groups 3–12) of the Periodic Table, have Catalytic Properties, high Melting Points and Densities. Aluminium has thirteen Protons and thirteen Electrons and is **Post Transition**, meaning it is in groups 13-15 on the Periodic Table.

Protons have a positive (+) charge, Neutrons have a neutral charge; both of these never leave the Nucleus. (**Just out of interest Protons and Neutrons are made up of sub-atomic particles called Quarks**) Electrons, however, in orbit around the Nucleus will have a negative (-) charge.

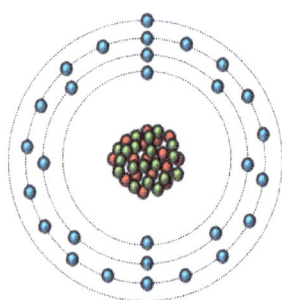

On the left is a Copper (Latin **Cu**prum Cu) Atom - Atomic Number 29. The number of Protons is always equal to the number of Electrons so our Copper Atom that has twenty-nine Protons will also have twenty-nine Electrons arranged in orbits and in this case there are four orbits. The orbits are called '**Shells**' and are arranged so the closest to the Nucleus will have up to two Electrons, the second orbit several Electrons. Other orbits can take many more and will make up the total number of Electrons for that element. The Electrons that are the easiest to move of course and are the '**loosest**' are the ones in the outer orbits or '**Valance**' Protons and Neutrons remain locked in the Nucleus of the Atom and do not move.

Sum up Questions on Atomic Structure:

Q1 – Are Electrons positively charged, negatively charged or neutral?

A1 – **NEGATIVELY CHARGED.** They are in orbits called 'shells' around the nucleus. Protons are positively charged and Neutrons, as the name suggests, are neutral charge.

Q2 – Can you get smaller particles than Electrons, Protons & Neutrons?

A2 – **YES.** Most books might say no, but the three mentioned particles are made up of very small particles which are called **'Quarks'**

Q3 – How are atoms donated in a formula?

A3 – **ATOMS ARE SHOWN AS SMALL NUMBERS OR NO NUMBERS.** So if we take a common formula that everyone knows of water which is H_2O this is telling us that one Molecule of water is made up of three Atoms in all; two Atoms of Hydrogen (H_2) and one Atom of Oxygen (O). Another one is Carbon Dioxide CO_2 which is again three Atoms; one Atom of Carbon (C) and two Atoms of Oxygen (O_2).

Q4 – What is a Molecule?

A4 – **A SMALL UNIT OF AN ELEMENT OR COMPOUND.** So when you see a formula which is shown as $2H_2SO_4$, which is Sulphuric Acid, this means that this Compound has two Molecules with seven Atoms; two Atoms of Hydrogen, one Atom of Sulphur and four Atoms of Oxygen.

Q5 – What holds the structure of an Atom together?

A5 – **COVALENT BONDING.**

Q6 – When an Element has an 'Atomic Number' what does that mean?

A6 – **THE NUMBER OF ELECTRONS & PROTONS.** The number of Electrons in the shell orbits will always equal the number of Protons in the Nucleus.

Q7 – How do we determine how many shells the Atom has got and how many Electrons in each shell?

A7 – **BY THE FORMULA $2(n)^2$.** This is a very complicated answer. In the formula 'n' is the shell number so in the first shell it would read $2(1)^2 = 2$ so there would be two Electrons in the first shell. In the second shell it would read $2(2)^2 = 8$ so there would be eight Electrons in the second shell and so on. That is as far as we will go here.

Q8 – What is an Atom's outer Electron orbit called?

A8 – **THE VALANCE.**

Q9 – What are 'Loose' Electrons?

A9 – **THE ONES IN THE OUTER SHELL OR VALANCE.** When a current flows in a conductor it is these Electrons that move the easiest.

Q10 – Why do Element letters on the Periodic Table sometimes not match the substance? For example Copper (CU), Potassium (K), Hg Mercury etc.

A10 – **OBSOLETE NAMES & LATIN.** For instance Copper is **Cu** because **Cuprum** is an obsolete name for Copper. Potassium **K** because **Kalium** is Latin for Potash. Mercury is **Hg** because of the Latin name **Hydrargyrum.**

What is a Conductor?

Again, before we go too much further, what makes a '**Conductor**' different from an '**Insulator**'? Let us firstly have a look at two popular metals that are very good conductors and are used extensively in the production of electric cables and current-carrying equipment.

Question:- What makes these '**Electrical Conductors** better than each other? **Answer:-** Their Atomic Structure.

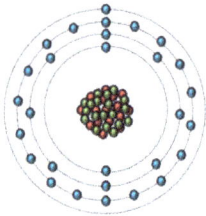

Copper

Chemical Symbol Cu

Atomic Number 29

Nucleus 29 Protons

Nucleus 34 Neutrons

4 Orbits of Electrons

29 Electrons

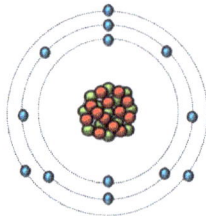

Aluminium

Chemical Symbol Al

Atomic Number 13

Nucleus 13 Protons

Nucleus 14 Neutrons

3 Orbits of Electrons

13 Electrons

The **Atomic Number** on the **Periodic Table** i.e. Copper 29 and Aluminium 13 refers to the number of Protons in the nucleus. The number of Protons as you will see in the diagrams on the left, is always equal to the number of Electrons. So our Copper Atom has twenty-nine Protons and going by the above it will have twenty-nine Electrons in four orbits. Aluminium thirteen Protons so will have thirteen Electrons arranged in three orbits and so on.

The orbits are called '**shells**' and are arranged so the closest to the nucleus, first shell, will have up to two Electrons; the second orbit, second shell; several electrons. Other orbits can take many more and will make up the total number of Electrons for that Element.

So if we now look at the atomic properties of these metals we find that the metallic bonds of the Electrons are very 'loose' and if a voltage is applied from, say, a battery they will freely move around the material from atom to atom, hence we have now got an electric current flowing. The Electrons that are the easiest to move of course and are the 'loosest' are the ones in the outer orbits or '**Valance**' which usually do not number many. The more loose Electrons in the outer or 'Valance' orbit the easier they move. Protons and Neutrons remain locked in the nucleus of the atom and do not move in the event of applying a voltage. When Electrons move they leave '**holes**' for others to move into.

Shells:

How do we determine how many shells an Atom has got? By the formula $2(n)^2$. This is a very complicated answer. In the formula 'n' is the shell number, so in the first shell it would read $2(1)^2 = 2$ so there would be two Electrons in the first shell. In the second shell it would read $2(2)^2 = 8$ so there would be eight Electrons in the second shell and so on.

Conductors:

The above metals or conductors are used extensively for electric cables in the electrical world. '**Copper**' is used in general cabling more than '**Aluminium**' which, although it is not used so much in wiring equipment in domestic and industrial premises, is used for high voltage from pylon to pylon on the national grid because it is much lighter than copper.

Metal:	Chemical Symbol:	Atomic Number:	Protons:	Neutrons:	Orbits:	Electrons:
Silver	Ag	47	47	60-62	5	47
Gold	Au	79	79	118	6	79
Zinc	Zn	30	30	35	4	30
Nickel	Ni	28	28	31	4	28
Brass	Alloy of Copper and Zinc					
Bronze	Alloy of Copper and mainly Tin					
Iron	Fe	26	26	30	4	26

The table above shows some **other** metals, besides Copper and Aluminium, which conduct electricity; best at the top to fair at the bottom. We do not make our cable conductors of Gold or Silver for obvious reasons although certain contacts of more expensive relays/contactors may have a Silver coating.

Sum up Questions on Conductors:

Q1 – What makes the Electrons move to make a current in a conductor?

A1 – **A VOLTAGE.** By applying a voltage, AC or DC to a conductor, will result in an electric current flow.

Q2 – Why do Conductors actually work?

A2 – **LOOSE ELECTRONS.** You will find that if we look at the atomic structure on the previous page and look at the Electrons in orbit around the Atom Nucleus, the Electrons in the outer orbit, or the valance, are easily moved from Atom to Atom when a voltage is applied and are called 'Loose' Electrons as their bonds to the Atom are not tight as they would be in an Insulator.

Q3 – What materials are the best conductors?

A3 – **METALS, CARBON ETC.** Most metals will conduct electricity – some better than others. You may recognise Copper and Aluminium in the electrical world as good conductors. It may be that Gold and Silver are even better, but expensive. Other materials will also conduct, for example Carbon, which is used for motor & generator brushes etc.

Q4 – What makes one conductor different from another?

A4 – **THEIR ATOMIC STRUCTURE.** If we take Copper and Aluminium for instance. Copper's Atomic Number on the Periodic Table is 29, which means of course that it has twenty-nine Electrons in orbits and twenty-nine Protons in the Nucleus.

Aluminium's Atomic Number is 13 which means it has thirteen Electrons in orbits and thirteen Protons in the Nucleus. (Previous section.)

Q5 – Is water a good conductor of electricity?

A5 – **NO.** It is not exactly a bad conductor, but because it will conduct and spreads all over is what makes it dangerous. If it was a good conductor we could have hosepipes full of water instead of cables, which would be much cheaper. Distilled water is, of course, an Insulator.

Q6 – Do conductors do their job better on AC or DC?

A6 – **EITHER.** DC is the most efficient voltage, but its problem is a phenomenon called 'Volt Drop' over long distances. Hence the National Grid is 450,000 Volts AC. Just out of interest they have just run a cable 450 Miles from Norway to Northumberland to power 1.4 million homes. It would have to be around 500,000 Volts AC otherwise if it was DC we probably could not be able to run a train set on this end! Another problem with DC is changing the voltage, as transformers cannot be used on DC.

Q7 – Are the conducting cables to Pylons on the National Grid made of Copper?

A7 – **NO!** Aluminium cables are used to go from pylon to pylon because Aluminium is less dense and so much lighter. Copper cables would be far too heavy.

Q8 – Are Industrial and domestic power and lighting cable conductors Copper?

A8 – **YES.** For industrial and domestic conducting circuits the wiring for lighting and most equipment would be Copper. **Aluminium cable cores cannot be used under 16mm² in a hazardous zoned area.**

Q9 – Is a Semiconductor a conductor or an insulator?

A9 – **BOTH.** It is an insulator until some change causes it to be a conductor.

What is a Semi-conductor:

A Semiconductor: is in actual fact an '**Insulator**' until certain conditions change and it becomes a '**Conductor**'. That could be a change in voltage, temperature, light or the addition of impurities called doping (such as Arsenic or Phosphorus). Conductors have the ability to have current flowing through them by having 'free' Electrons in an outer orbit so by doping we add an Element that has 'free' Electrons to our semiconductor compound with no 'free' Electrons to encourage it to conduct, i.e. the free Electrons from the doping will move and hence provide an electric current when certain conditions apply as above.

Doping:

Arsenic & Phosphorous would give '**N**' type doping as the free electrons would be a Negative charge. Boron for instance would be a '**P**' type doping, this encourages '**holes**' in the **Covalent Bond** (below) which would leave the Electrons floating around with nothing to bond to, so would give a positive charge.

Zener Diodes are a good example of a Semiconductor that changes with voltage. The Zener Barrier that contains the Zener Diodes protects the Intrinsically Safe equipment in the zoned area from a dangerous rise in voltage. One lamp of course that uses this technology is the **LED Lamp.** Integrated Circuits and Photovoltaic Cells (Solar Cells) are another abundant use of semiconductors.

Two Semiconductors that readily spring to mind are Silicon and Germanium, but these emit their energy in heat and not light. Other semiconductors which emit their energy in the form of light are used in LED lamps and these are Gallium Arsenide (GaAs), Gallium Phosphide (GaP) and Gallium Arsenide Phosphide (GaAsP) depending upon what colour LED lamp is required.

Silicon (Si):

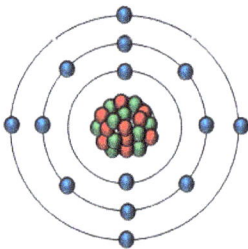

Let us first look at Silicon; this compound as you can see from the atomic diagram on the left has fourteen Electrons in three shells, which means that it will have fourteen Protons in its nucleus. Silicon has an operating temperature of above 100 degrees centigrade, which makes it attractive for many uses where temperature is a problem. The Electrons in the orbit shells are in what is called Covalent Bonds with neighbouring Electrons in their Atoms so unlike metals they are not free to just move around and form a sort of lattice.

Germanium (Ge):

Now let us first look at Germanium; this compound as you can see from the atomic diagram on the right has thirty-two Electrons in four shells, which means that it will have thirty-two Protons in its nucleus. Germanium has an operating temperature of above 70 degrees centigrade which, although is lower than Silicon, makes it attractive for many uses where temperature is a problem. The Electrons in the orbit shells are in Covalent Bonds with neighbouring Electrons in their Atoms so unlike metals they are not free to just move around and form like a lattice.

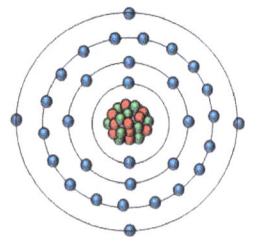

So a Semiconductor is a compound insulator, usually of Silicon or Germanium, whose Atoms do not have 'free' electrons, and to make it a conductor can be done by:-

a) 'Doping' i.e. by adding Arsenic or Phosphorous to add free Electrons and create an 'N' type (Negative), or something like Boron to create 'holes' in the Covalent Bonding which would be a 'P' type (Positive). Doping the Semiconductor makes them called '**Extrinsic**' and other means of Electron movement is called '**Intrinsic**' as the next actions are called.

b) A breakdown voltage can be inserted to make the Electrons move out of their bonding.

c) Light to cause enough EMF for Electron movement.

Without this technology there would be no electronic items such as Transistors, Diodes or in fact any Integrated Circuits so Semiconductors play a very important part in our electronic technology.

Sum up Questions on Semi-conductors:

Q1 – Is a Semiconductor an insulator or conductor?

A1 – BOTH. It is an insulator until something else is done to make it a conductor, i.e. Doping, Breakdown Voltages etc.

Q2 – If the doping was Arsenic or Phosphorous what type would this doping be?

A2 – 'N' TYPE. To install **'free'** electrons into the Semiconductor which would give it a negative charge.

Q3 – If the doping was Boron what type would this doping be?

A3 – 'P' TYPE. To install 'holes' in the **'lattice'** which would give it a positive charge.

Q4 – What are the most common Semiconductors?

A4 – SILICON AND GERMANIUM. Of which I am sure you may have heard?

Q5 – Are Semiconductors used in lighting?

A5 – YES. They are used in LED lamps. The Gallium range of Semiconductors, **NOT** Silicon or Germanium in this case.

Q6 – What are 'Extrinsic' Semiconductors?

A6 – DOPED. These are called **'Extrinsic';** others like excess voltage or light are called **'Intrinsic'** as below.

Q7 – What are **'Intrinsic'** Semiconductors?

A7 – OTHER MEANS THAN DOPED. Those which might be changed by voltage, light etc.

Q8 – Can Semiconductors operate in high temperatures?

A8 – YES. Silicon can operate at the temperature over 100ºC and Germanium at around 70ºC.

Q9 – What do they call the Atomic 'Bonding' of Semiconductors?

A9 – COVELANT BONDING. A sort of **'lattice'** where there are no 'Free' Electrons until the Semiconductor is doped.

Q10 – What does Silicon's Atomic Number 14 mean?

A10 – 14 ELECTRONS. In the shells around the nucleus, which means there are fourteen Protons that remain in the nucleus. This goes for any Element or Compound on the Periodic Table.

Q11 – Can the Semiconductor be made to turn on and off?

A11 – YES. The voltage and light operated Semiconductors will do this all of the time.

Q12 – Where are Semiconductors most used?

A12 – INTEGRATED CIRCUITS. And most electronic components i.e. Transistors, Diodes etc. Also used in Intrinsically Safe protection such as Zener Barriers.

What is an Insulator?

We looked previously at popular conductors of electricity. Now let us look at **'Insulators'**. **Question:-** What makes one material an 'Insulator' and another a 'Conductor'? **Answer:-** Their Atomic Structure.

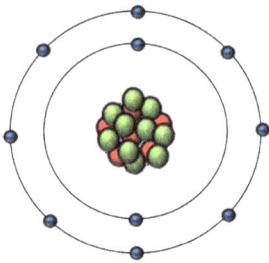

The number of Protons as you will see in the diagram on the left, is always equal to the number of Electrons. So our particular insulator atom has ten Protons and going by the above it will have ten Electrons and in this case we have two orbits or shells.

The orbits are called **'shells'** and are arranged so the closest to the nucleus in our case has up to two Electrons; the outer orbit or in our case 'Valance' has eight Electrons. Different materials will have different numbers of orbits or shells and a different number of Electrons and Protons

Shells:

So if we now look at the Atomic properties of these insulators we find that the bonds of the Electrons are very **'tight knit'** and if a voltage is applied from, say, a battery there are no 'loose' electrons so they will **NOT** freely move around the material from Atom to Atom, so there will be no electric current flowing.

In conductors, usualy metals, as mentioned, the electrons that are the easiest to move of course and are the 'loosest' are the ones in the outer orbits or valance which usually do not number many. Here the Electrons in the Valance do not move. Protons and Neutrons, as in the case of the conductor, remain locked in the nucleus of the atom and do not move at all!

Several electric cable insulations are listed in the table to the right. Some you may recognise straight away. Probably PVC is the most common cable insulation as the grey twin and earth in houses would be of this insulation. Sometimes the outer sheath may be a different material to that which covers the conductors.

Initials:	Full Name:
PVC	Polyvinyl Chloride
EPL	Ethylene Propylene Rubber
XLPE	Cross Linked Polyethylene
CPE	Chlorinated Polyethylene
PTFE	Polytetrafluoroethylene

Examples of Insulators:

Please also remember that insulation can break down for many reasons i.e. dampness & water. Water is **NOT** a very good conductor of electricity otherwise we could have hosepipes full of water for carrying our electricity which would be much cheaper than Copper. The fact is that water **WILL** conduct and speads all over which makes it dangerous. High voltage can also break down an insultation if high enough.

Paper	Wax
Wood	Tufnell
Nylon	Paper
Plastic	Distilled Water
Glass	Porcelain
Teflon	Oil

On the left are several more insulators, many of which you will recognise and most of these are used in the world of electricity in either equipment, batteries or cables.

As mentioned above the solid Insulators on the left only do their job correctly if they are **DRY!** Get them damp and they could become conductors.

Who discovered Electrons, Protons & Neutrons:

1) Electron: Negative Charge, Discovered in 1897 by J. J. Thompson

2) Proton: Positive Charge. Theorised in 1815 by William Prout.

3) Neutron: No Charge. Discovered in 1932 by James Chadwick.

Sum up Questions on Insulators:

Q1 – What makes the Electrons in an insulator different to a conductor?

A1 – **THE BONDING.** The bonds of the Electrons are very **'tight knit'** and if a voltage is applied from, say, a battery, they will **NOT** freely move around the material from Atom to Atom, so there will be no electric current flowing, so no 'free' Electrons in a valance.

High voltage will, of course always break down an insulator. For example on **SOME** electrical screwdrivers there is a warning for it not to be used on live systems over a certain voltage.

Sometimes this may be an end to end flashover of the high voltage arc rather than the insulator actually breaking down.

Q2 – What materials are the best Insulators?

A2 – **NON-CONDUCTORS.** PVC, plastic, rubber, wood, paper, glass, nylon, plastic, Teflon, distilled water, porcelain, vitrite & oil. Some of these materials are used in the manufacture of lamps, lamp holders and wiring. (Glass, vitrite, porcelain, plastic.)

All of these examples will become conductors if they become damp or the voltage gets too high.

Q3 – What makes one Insulator different from another?

A3 – **THEIR ATOMIC STRUCTURE.** The number of Electrons, Protons and Neutrons. Have a look at insulators on the Periodic Table.

Q4 – Do insulators do their job better on AC or DC?

A4 – **EITHER.** Both AC & DC equipment require insulation of one kind or another if we take electric wiring, the insulation on the outside of the wire is the same be it AC or DC.

Q5 – Why is distilled and deionised water an insulator?

A5 – **NO IMPURITIES OR CONTAMINANTS.** It is the impurities in water such as salts etc. that makes it a conductor although not a very good one. Adding salt will dramatically make the water a conductor so just think about seawater.

Q6 – Is air a good insulator?

A6 – **YES:** Otherwise switches, circuit breakers and many other electric/electronic devices would not work. I must stress that the air must be **DRY.**

Q7 – What is dielectric strength?

A7 – **BREAKDOWN VOLTAGE:** Put simply, the highest voltage known to break down a particular insulating material.

Q8 – What is the best insulator ever?

A8 – **POSSIBLY A PERFECT VACUUM:** This contains no impurities at all. Its dielectric strength will be enormous, hence Vacuum Contactors and Circuit Breakers.

Q9 – If air is an insulator and voltage can jump, how far will 240 volts **RMS** jump?

A9 – **IT WILL NOT.** At this low voltage you would actually have to touch it to get a shock.

What is Electron Flow?

Electron flow: occurs in two entirely different types of electricity, namely AC (**A**lternating **C**urrent) which comes out of say a 13Amp socket in your house and DC (**D**irect **C**urrent) that would be a battery voltage. DC is the most efficient of the two!

So we have decided that it is Electrons that move when a voltage is put onto a conductor so it is these particles that interest us most. Electrons have a negative (-) charge, Protons have a positive (+) charge and Neutrons have no charge. Electrons move and Protons and Neutrons remain stable in the Nucleus of the Atom.

Let us look at a typical copper conductor:

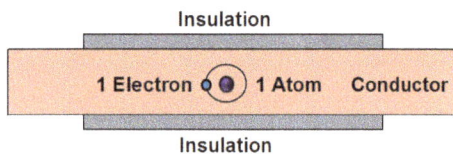

Left is a conductor, which let us say is the Copper conductor of an electric cable. The grey is the insulation. I have, for ease of explanation, only drawn one atom with one Electron. Let us put a DC voltage onto the conductor and see what happens.

Looking at the diagram on the right by putting a DC voltage onto the conductor you can see that the Electron, being negatively charged, has now moved from the atom towards the positive side of the conductor. So as we mentioned the Copper conductor is made up of billions of Atoms **ALL** with **'Free'** Electrons so they will **ALL** move in the same direction when the voltage is applied. So now we have Electron flow or current flow which is described below:

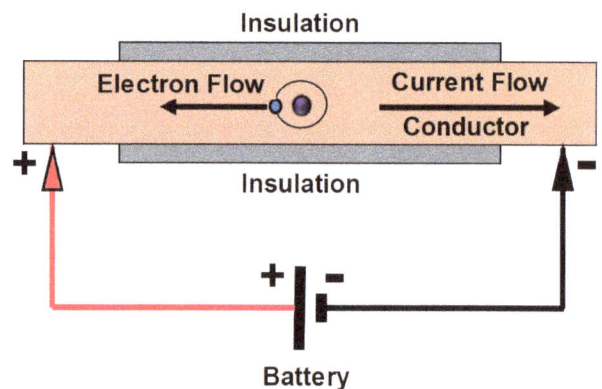

I am now going to complicate things a little. **Experiments** by **Benjamin Franklin** indicated that the current flow is in the opposite direction to the Electron flow. Now remember this is based on a theory by a great man and to tell you my opinion **'it really does not matter.'** How much current flows depends upon the voltage and what is called the 'resistance', in Ohms (Ω), of our conductor.

We mentioned earlier voltage (V), amps (A) and resistance (Ω):

Amps:

Current flow is measured in amps (A) so who invented these three readings? The amp, short for **Ampere**, is named after **André-Marie Ampère** who was a French Physicist and Mathematician. If we looked at an equivalent water system then we could actually for our analogy take the water flow as the amperage.

Volts:

We have mentioned volts and they are named after **Alessandro Volta** an Italian Physicist. So in our analogy the pressure behind the water and the force that is pushing it through the piping would be the voltage.

Ohms:

The resistance of the conductor would be measured in Ohms (Ω) named after **Georg Simon Ohm** who was a German Physicist. Now here in our analogy we could say that any restrictions in the pipework i.e. the pipe narrowing or going through a filter etc. would be our resistance.

Electron:

Discovered in 1879 by **J. J. Thompson**.

Sum up Questions on Electron Flow:

Q1 – Is it volts or amps that actually flow?

A1 – **CURRENT (AMPS).** Flows Positive to Negative.

Q2 – Is Current flow and Electron flow the same thing?

A2 – **NO.** According to Franklin. Electron flow is the opposite to the Current flow. Electron flow is from Negative to Positive.

Q3 – What initiates current flow?

A3 – **VOLTAGE.** If we take an analogy of water then the water flow is the current and the water pressure is the voltage.

Q4 – What Initiates Electron flow?

A4 – **EMF (VOLTAGE).** Electromotive Force.

Q5 – Are Electrons Positive or Negative charged?

A5 – **NEGATIVE.** Electrons have a Negative charge, Protons have a Positive charge and Neutrons have a Neutral charge.

Q6 – When was the Electron discovered and who by?

A6 – **J. J. THOMPSON IN 1879.** When he was working on a Cathode Ray tube.

Q7 – Do we get Electron flow in AC or DC?

A7 – **BOTH.** But in DC they only travel in one direction.

Q8 – Can Electrons flow without a voltage?

A8 – **NO.** It is obviously possible to have a voltage without electron or current flow, but it is not possible in a standard circuit to have Electron or current flow without a voltage.

I have actually stated **'NO'** but in theory 'Free' Electrons that are in the outer valance are constantly on the move as they circle their Atoms and maybe even filling random holes from Atom to Atom, but by putting a power source onto them they all move in the same direction along with the current.

Q9 – What are 'Excited State' Electrons?

A9 – **ELECTRONS THAT JUMP ORBITS.** Which they do from inner orbits when they absorb energy.

Q10 – Can magnetism cause electrons to flow?

A10 – **YES.** Motors and generators use magnetic fields to work. By moving a magnet in and out of a coil can cause an EMF and this would be called 'Electromagnetic Induction'.

Emil Lenz conducted some experiments on this subject. Induction Lamps also use magnetism around their Induction Coils.

Q11 – Have Electrons got any mass?

A11 – **VERY LITTLE.** Which is why they move easily when energy is applied.

AC Voltage:

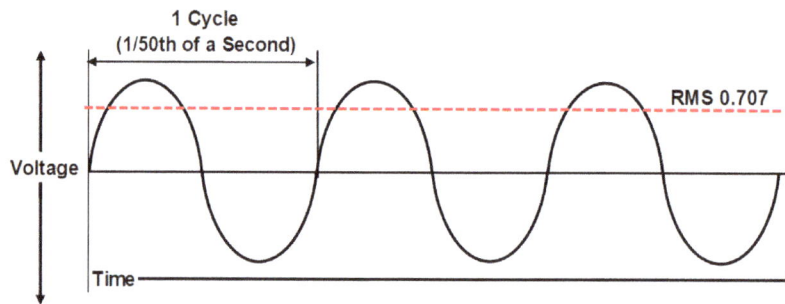

Frequency:

The diagram above shows a number of cycles in the UK and the frequency of the electricity would be 50 Cycles per Second (CpS) or 50 Hertz (Hz) as we call it now. In the USA this would be 60Hz. So each cycle of AC electricity would go up to positive peak and down through zero to negative peak which we know as a sine wave. AC voltages can easily be changed using devices called transformers, something that would be very difficult with DC as transformers only work with AC.

Just let us look at DC for a moment. Here there is just a positive (+) and a negative (-) no matter how large or small the voltage is, but with AC that is quite different as there are 3 phases, each 120 degrees out of phase with the other two. The colours used to be red, yellow & blue, but now they are brown, black and grey. Many industrial items of equipment are three phase including the majority of electric motors. Domestic houses use one of these phases, which is called single phase, and the wiring is live (brown) and neutral (blue).

AC voltage is used in our homes and is transmitted in very high form (around 450,000 volts) all over the country by the national grid and transformed down through substations. AC is used by the grid as a voltage that can be transmitted many miles without much volt drop. As mentioned earlier, the longest cable run in history, around 450 miles from Norway, where they use hydro (water) power, to Northumberland in the UK, has just been opened which will obviously be AC as if it was DC, with volt drop, there would be very little power left by the time it got to the UK.

Power Factor & Unity:

I have put a dotted red line on the diagram above, which is the **Root Mean Squared Value or 0.707.** This is the most efficient or effective part of the sine wave or the DC equivalent section. When test instruments are put onto a circuit they read this **RMS** value not the peak value. Even electricity meters read this value. This could also be called the DC equivalent voltage of an AC Sine Wave.

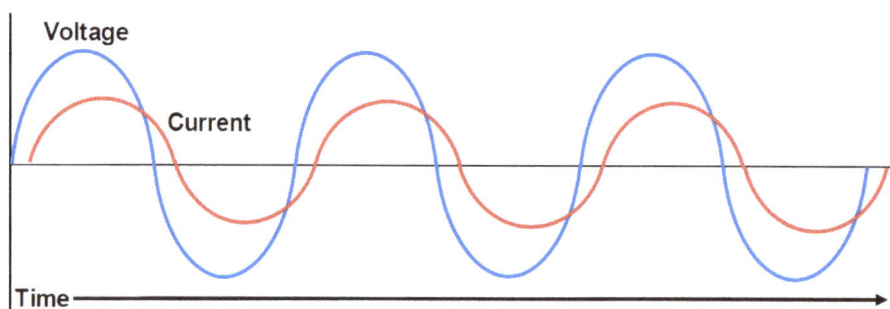

I often get asked if the sine wave is voltage, what has happened to the current? Well looking at the diagram above, the blue sine wave is voltage and red sine wave is current. You will notice that the red sine wave is lagging the voltage which will tell us that this is an **'inductive'** load such as a coil. If we got a capacitive load then the current would lead the voltage. The cosine of the angle by which the current lags or leads the voltage is what is called the **'Power Factor'** which suppliers try to keep to as near **'Unity' (1)** as possible. Usually industrial power factors run at around 0.8.

Sum up Questions on AC Voltage:

Q1 – What is the frequency of an AC Voltage?

A1 – **50 HERTZ.** In this country the frequency, or Cycles per Second as it used to be called is now 50Hz. We are talking about one cycle being 1/50th of a second. It is demonstrated on any drawings as a Sine Wave, but the waves can be all sorts of shapes from triangular to square which may be, basically, what an inverter produces before other equipment is used to make them roundish. In the US this could be 60Hz.

Q2 – Why do we use AC?

A2 – **EASIER TO TRANSMIT & TRANSFORM.** AC can be transmitted over very long distances and can be transformed up and down quite easily. Transformers will of course not work on DC so controlling it would be extremely difficult and any distance would cause volt drop. As stated earlier, a cable has been run approximately 450 Miles from Norway to Northumberland to power 1.4 million homes. It would have to be around 500,000 Volts AC otherwise if it was DC we would probably not be able to run a train set on this end!

Q3 – Do instruments read the peak voltage at the top of the Sine Wave?

A3 – **NO.** All instruments from Voltmeters to your Electric Meter at home read the RMS (Root Mean Squared) voltage, which is 0.707 of the peak.

A4 – How many phases are there?

Q4 – **THREE.** We actually generate three phase, 120 degrees apart making 360, which is transformed up and down by transformers in substations. Used to be Red, Yellow & Blue now Brown, Black & Grey. It is transmitted as three phase throughout the country at various voltages the largest being around 450,000Volts. Industry will use the three phases in various items of equipment i.e. Electric Motors throughout the factory, but domestic premises will usually only use one of the phases plus a neutral which is created at the distribution transformer. (Brown and blue.)

Q5 – We have talked about Voltage, where is the Current?

A5 – **USUALLY LAGGING THE VOLTAGE.** With an inductive system, i.e. transformers and motors with so many coils of wire, the current is another smaller Sine Wave lagging the Voltage Sine Wave. If the system is capacitive then the current Sine Wave will lead the Voltage Sine Wave.

Usually we talk about inductive systems. The Cosine of the angle by which the current lags or leads the voltage is called the **Power Factor (Cos φ Phi)** which should be as near to **1 (Unity)** as possible. We can correct the power factor by inserting huge Inductors or Capacitors into the system. Industrial Power Factors may run at around 0.8.

Q6 – How can we change electric motor speed or lighting brightness?

A6 – **CHOP THE SINE WAVE OR CHANGE FREQUENCY.** We used to do this task by varying the voltage, but with motors this could result in reduced torque, and dropping the voltage causes a problem with discharge lamps, which includes fluorescents. We can do it this these days with **Pulse Width Modulation.**

Q7 – Can we measure AC voltage?

A7 – **YES & NO. ALL** instruments measure current and change to voltage on the scale for you.

DC Voltage:

DC Voltage: is the most efficiant when comparing it with an AC voltage. One of the main drawbacks is volt drop. DC cannot be transmitted over large distances like the national grid. The longest ever cable run is from Norway to Northumberland in the UK at around 450 miles long. It will power thousands of homes with green energy as Norway is 'hydro' (Water) power. If this cable run was DC it would not work. There are several types of DC let us have a look at some of them and this may give you the idea.

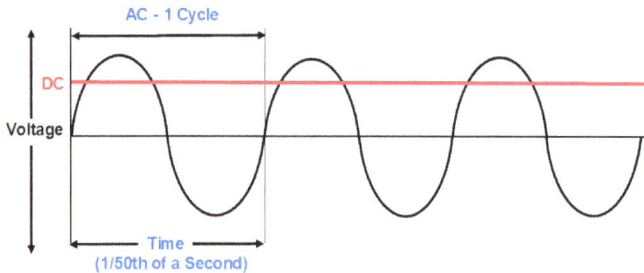

Changing from AC to DC is fairly easy. I have shown AC sine waves (left) and the DC in red. I have shown the DC as 0.707 of the sine wave which is the 'Root Mean Squared' (RMS) value (explained in another section). So the straight line of DC is what we are after as our most efficient voltage so how do I obtain it?

AC to DC:

Well I could get a battery, the output of which would be DC straight line similar to the red line above and I could not have a better example. The voltage would depend upon my battery. So if I want to create DC another way, say, from AC then that would be a little bit more involved, so how?

If I take the AC sine wave above and put a diode into the supply, as the AC voltage goes first positive, then drops down to negative and then positive again, putting the diode in would stop the sine wave from going negative hence I would have a type of DC be it very 'noisy' as per the diagram on the right. This would be 'half wave'.

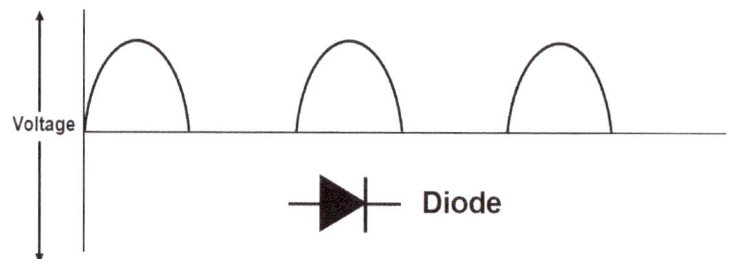

What do I mean by 'noisy'? There are gaps in the distance between the peaks so there would be a peak every 1/50th of a second and then a gap before the second one etc.

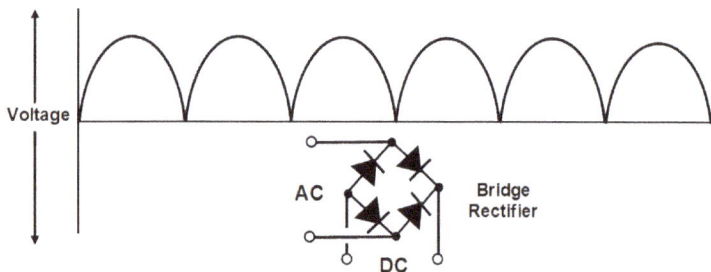

By using a bridge rectifier shown left I can in fact invert the negative part of the sine wave up to the positive to fill in the gaps which would be less 'noisy'. As you can see we are now getting closer to a straight line which would, as in the case of the battery, be our ideal DC output. This would be 'full wave'.

So can we improve on the above to make our DC even smoother? The answer is yes as I have shown below.

The top red line is our battery. By adding a capacitor to the circuit I can fill in the valleys and this may be as good as I can achieve. This DC would be quite smooth and could be used for my DC equipment including instruments. I hope that this page has shown how we can use rectification and capacitors to obtain our DC.

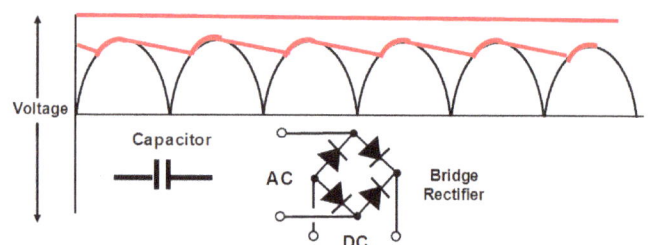

Sum up Questions on DC Voltage:

Q1 – Is DC Voltage more efficient than AC Voltage?

A1 – YES. In AC how can a voltage be efficient if it is constantly passing through zero?

Q2 – Are there phases with DC?

A2 – NO. All one system similar to single phase in AC.

Q3 – Can I change DC to AC?

A3 – YES. With a piece of equipment called an Inverter.

Q4 – Can I change AC to DC?

A4 – YES. We do this easily and quite regularly with our battery chargers. Generators that produce AC can be fitted with a commutator to change output to DC.

Q5 – Will Transformers work on DC?

A5 – NO. Transformers need the induced voltage of the primary to cutting the coils in the secondary at 50Hz to enable them to work. DC voltage can be dropped using equipment such as a rheostat.

Q6 – Will AC electric lamps work on DC?

A6 – CERTAIN ONES. If they are incandescent lamps and car or torch bulbs, i.e. they have a metal filament. If they are discharge lamps then NO; remember that here we have a ballast controlling the current in an arc tube.

If we look at AC equipment plugged into DC that is totally different. Because there is no frequency there is no inductive reactance so it may see the AC equipment with its coils as a short circuit and burn out. It is possible to obtain universal equipment for AC or DC (Universal Motor).

Q7 – Why not use DC instead of AC?

A7 – AC IS EASIER TO TRANSMIT & TRANSFORM. AC can be transmitted over very long distances and can be transformed up and down quite easily. Transformers will, of course, not work on DC so controlling it would be extremely difficult and any distance would cause volt drop. DC motors are used on equipment such as cranes as we have much more direct control. DC is used on car electrical systems, children's electric toys, torches etc. courtesy of the battery. DC is used on most electronic circuits.

Q8 – Is there such a thing as a Power Factor with DC?

A8 – NO. There is no frequency involved with DC so the current will always be in phase with the voltage. I have said 'NO' but we could say that the power factor is always unity with DC as current always in phase with the voltage.

Q9 – Is Static Electricity DC?

A9 – YES. Including lightning, which of course, is very high voltage Static Electricity.

Q10 – Can I measure voltage?

A10 – NO. ALL instruments measure current and change to voltage on the scale for you.

The Multi-meter Section:

This section discusses the different sections of a Multi-meter and how it can be used to test various components of a printed circuit. We discuss the benefits of using a multi-meter with certain facilities.

Remember that if a multi-meter is used to test Intrinsically Safe Loops or certified Intrinsically Safe Equipment then the meter must be 'APPROVED' for working on certified IS loops and equipment.

I have drawn a diagram below of a typical multi-meter with a whole range of facilities for testing many different electronic components.

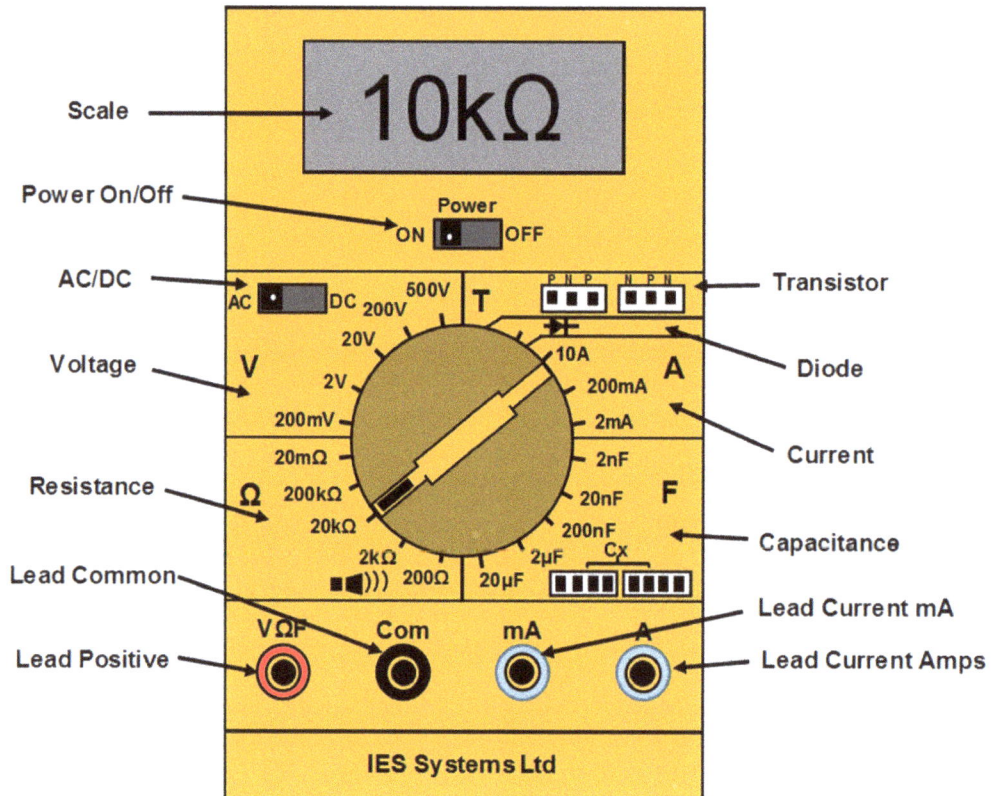

As we go through this section in more detail we will discuss how all of the different scales of the multi-meter are used and what the advantages are of using one against another.

These instruments require calibration now and then to ensure that they are still as accurate as possible; after all if the test instrument is not accurate what are you using it for? Also remember that the leads require a calibration check to ensure that they have no broken wires, open circuits or high resistances otherwise it can again prevent accurate readings.

Safety of the Meter and leads is also discussed such as the GS38 shields that protect the user. Also the dangers of using it on voltage whilst set on the wrong scale.

Usually more expensive Multi-meters have more of a selection of facilities for testing a wide range of electronic components.

The more expensive meters read 'True' RMS Voltage (TRMS) and are much more accurate. The RMS value only applies if the Sine Wave being measured is perfect. Sometimes electrical noise is present and the Sine Wave curvature is imperfect so this is where TRMS applies.

Multi-meters should not be used as primary potential indicators. Remember they also require a Gas Free Certificate in a Zoned Area.

The Multi-meter:

General:

Multi-meters are very handy instruments; many can be selected to read Volts, Amps, Resistance, Capacitance and more. They do of course read as any instrument does, RMS (Root Mean Squared – 0.707) and not peak Voltage. Only on perfect Sine Waves such as the mains.

These days Sine Waves, in the world of Electronics and Computers, are not forced to be perfect so another value comes into play i.e. TRMS. Expensive multi-meters read 'True' RMS (TRMS) and not just 0.707.

When being used on Atex Intrinsically Safe Circuits or Components then the Multi-meter must be APPROVED for working on IS Circuits and Components! Gas Free Certificate in Zoned Areas.

Analogue or Digital:

Most Multi-meters these days are digital as shown in the diagram below. Older Multi-meters similar to the old 'AVO' were analogue moving coil for accuracy. The expensive models even had a mirror at the back of the scale so that when the reflection of the pointer was in line with the actual pointer then that was when the most accurate reading could be obtained.

Digital meters are more compact and smaller which is handier to carry and also to use. These days the meters have a much larger range of testing than older meters.

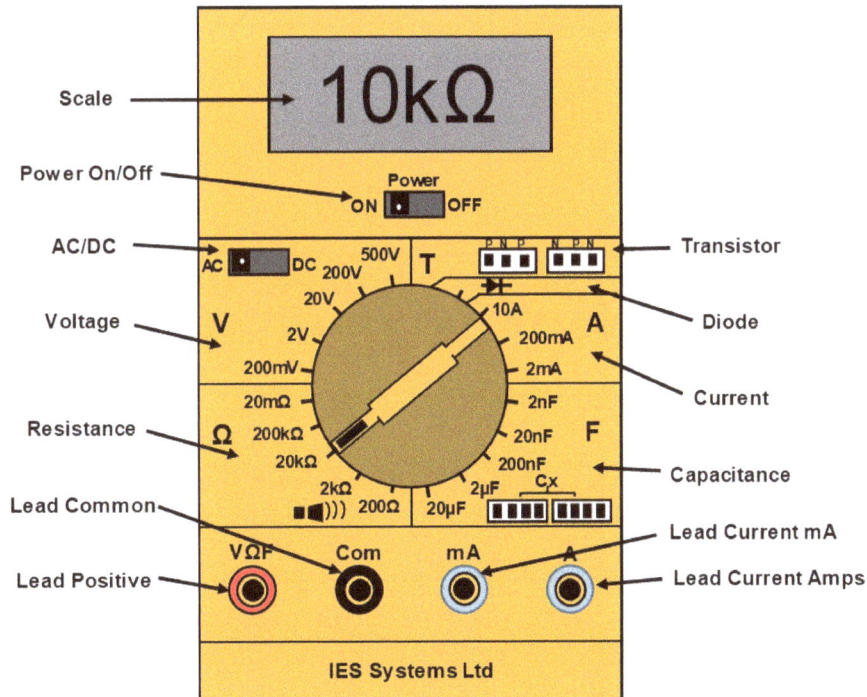

Safety:

Multi-meters are designed for test purposes only and should never be left connected into a circuit as a permanent fixture. They should not be used as a Primary Potential Indicator or on high energy circuits.

Instrument Internal Power:

The internal power for tests such as Resistance (Ω) comes, usually, from a 9 Volt PP3 Battery. Only approved Manufacturer advised batteries should be used.

CAT Ratings:

Always check the CAT Rating of your Multi-meter. They go from CAT ll (Electronics) to CAT lV (Outdoor Lines & Distribution). The meter should have protection against incorrect use.

Voltage (V): Red lead in red terminal and black lead in common. These readings can be obtained using the leads. You would have to select 'AC' or 'DC' on the selector switch when testing voltage.

Resistance (Ω): Red lead in red terminal and black lead in common. These readings can be obtained using the leads. Reading would be in Ohms. Continuity may have a sound (Beep) facility. The test voltage is usually from a 9 Volt PP3 Battery; do not use on equipment that cannot stand this voltage!

Capacitance (F): Red lead in red terminal and black lead in common. These readings can be obtained using the leads. Reading would be in Farads. If the multi-meter has this facility Capacitors can be tested using the Cx test positions on the meter.

Current (A): Has its own plug in terminals which may include using the common. Current in Amps of course will be tested with the meter in series.

Transistors: On many instruments there is a facility to test a **PNP** or **NPN** Transistor using the test positions on the meter.

Diodes: Some multi-meters have a Diode Test Facility with a sign of a Diode. This is better than testing diodes using resistance (Ω) as it measures the Volt drop in forward bias which is more accurate. The reading would be in Volts not Ohms.

Scale Safety:

IT IS IMPERATIVE THAT THE METER IS NOT LEFT SELECTED TO READING THE WRONG SCALE WHEN TESTING. FOR EXAMPLE, IT MUST NOT BE SET ON RESISTANCE (Ω) OR CURRENT (A) WHEN TESTING FOR VOLTAGE (V) AS DAMAGE/EXPLOSION IS POSSIBLE!

Some Multi-meters have a cut out and some DO NOT.

Testing Intrinsically Safe Loops or Equipment:

If you are using the Multi-meter to carry out testing on Atex **Intrinsically Safe Equipment** the Meter will need to be 'APPROVED FOR WORKING' on Intrinsically Safe Equipment, but **not Intrinsically Safe itself**. If you are in a Hazardous area a Gas free Certificate would have to be obtained to use the meter.

Test Lead Connection to the Multi-meter:

Multi-meters have a red lead and a black lead. Unless you are measuring current, the **BLACK** lead plugs into the black terminal on the meter marked **'COMMON'**; the **RED** lead plugs into the terminal marked something like **'VΩ'**. Amps are usually separate terminals altogether.

Test Lead Safety:

1) Check the leads before you start any testing to ensure that they are in good condition with no metal conductor showing through the insulation.

2) Always use the correct leads for the Multi-meter not home-made, modified or damaged ones.

3) When testing always keep your fingers behind the GS38 test probe shields for safety. Holding the probes near to the tip can be dangerous. Do not use on high energy circuits

Checking & Testing the Leads:

A good test might be to set the meter onto Ohms (Ω) and touch the tips together. A low Ohms (Ω) reading may indicate the continuity of the leads. Whilst the tips are touching move the leads about to see if the reading changes. You are looking for a reading for the leads alone of around 0.4Ω.

It is handy to store the leads inside the multi-meter case on completion of testing, but constantly coiling up the leads to fit into the case can cause problems so they must be checked regularly.

Multi-meter Questions and Answers:

Q1 – Does the Multi-meter read peak voltage?

A1 – NO. Usually RMS value which is 0.707 of peak. Cheaper ones may even read average voltage 0.64. Many expensive types these days read 'True' RMS (TRMS) which is much more accurate if the Sine Wave being measured is not perfect and contains electrical noise.

Q2 – Can a Multi-meter read AC and DC Voltage?

A2 – YES. On most meters it is just a case of flicking a switch. On others there may be two separate sections for AC & DC.

Q3 – Does it matter which way round the leads are connected?

A3 – YES. On a polarity conscious component such as a Transistor or a Battery, Black is plugged into common. On some meters you may get a minus sign if they are wrong polarity.

Q4 – Can you overload a Multi-meter?

A4 – YES. But it will read O/L if you do. It may operate its trip or fuse if the overload is large.

Q5 – Can being on the wrong scale be dangerous?

A5 – MOST CERTAINLY. e.g. leaving the instrument set to OHMS and putting it on Volts.

Q6 – If the Multi-meter is used on Atex Intrinsically Safe Equipment does the actual meter have to be Intrinsically Safe?

A6 – NO. It has to be APPROVED for work on Atex IS Equipment. If the meter is not Atex itself then a Gas Free Certificate will be required for the meter.

Q7 – Is there a special position on the Multi-meter for testing Diodes & Transistors etc.?

A7 – SOMETIMES. On more expensive Multi-meters it looks like a Diode Icon.

Q8 – Are the Multi-meter leads fused?

A8 – NO NOT USUALLY. Be careful using it on high energy circuits. Remember the GS38 shields. They are not to be used as test lamps.

Q9 – Is an Analogue Multi-meter moving iron or moving coil?

A9 – MOVING COIL. On expensive meters for more accuracy.

Q10 – Can a Multi-meter be used to test high voltage such as Distribution 3.3KV & 6.6KV?

A10 – DEFINATELY NOT. This would be extremely dangerous to life!

Q11 – What does 'hFE' mean on a Multi-meter?

A11 – Hybrid Parameter Forward Current Gain. When testing transistors it is the measure of the junction DC current gain (Amplification). Transistors may require this test.

Q12 – Do Multi-meters have internal fuses or resets?

A12 – SOME OF THEM.

Q13 – Can I measure Frequency on a Multi-meter?

A14 – SOME OF THEM. The selector would have to have a position marked Hz.

The Printed Circuit Board Section:

Printed Circuits are circuits that manufacturers put into their appliances and they are specific to that piece of equipment, not a generalisation like Vero board which you might use in a hobby. Some good examples might be most equipment using electronics such as Atex Instruments and Systems on Chemical Factories and Platforms as well as laptop, PC, television, fridge, washing machine or radio circuits etc.

As we go on through the book we will look at the printed circuit and all of the different components that you might find on it and how Instrument and Electrical Technicians might remove them from the circuit without damage. Components of Atex Instruments are subject to manufacturer approval, see note at the bottom.

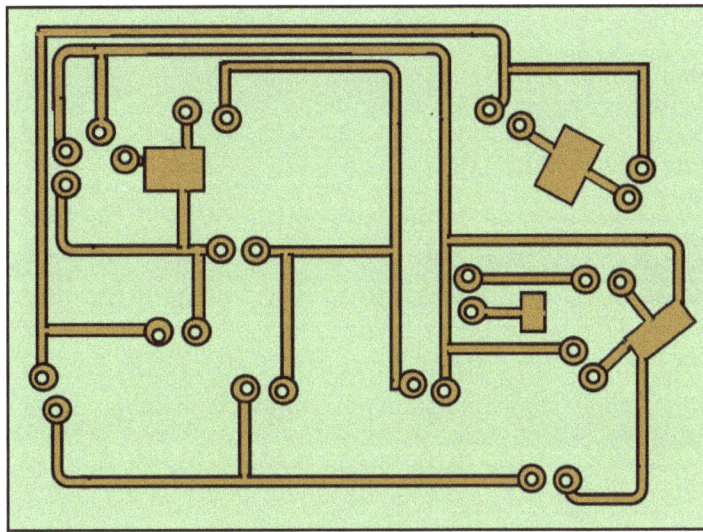

We discuss in this section how a board is made up, the materials used in their manufacture and how the track is put on.

The diagram above shows a **'Through Hole'** type PCB but there are other types such as surface mount.

We talk about a 'Breadboard' later, which is a board that can accommodate an electronic circuit for design or testing before soldering. Components can be tested individually on this board. It gets its name from the past where breadboards, believe it or not, were used for testing circuits by installing screws into the board and running wires to them!

NOTE:

THIS BOOK IS ONLY MEANT TO GUIDE YOU INTO FITTING/REMOVING COMPONENTS & FINDING A FAULT ON AN ELECTRONIC PRINTED CIRCUIT.

REPAIRS AND REPLACEMENTS TO PRINTED CIRCUITS ON ATEX CERTIFIED EQUIPMENT FOR HAZARDOUS AREAS, SUCH AS ZONES, CAN ONLY BE DONE WITH MANUFACTURER APPROVAL IN WRITING.

FAILURE TO COMPLY WITH THIS MAY BE CLASSED AS AN UNAUTHORISED MODIFICATION AND MAY RENDER ANY CERTIFICATION DOCUMENTATION NULL AND VOID!

Printed Circuits:

What are Printed Circuit Boards?

Printed Circuits are circuits that manufacturers put into their appliances and they are specific to that piece of equipment, not a generalisation like Vero-board that you might use in a hobby. Some good examples might be a television or radio circuits.

The metal interconnecting sections are called **'tracks'** and connect one component to another to make up the circuit. It is a type of miniature 'Bus' system. The finished product with all of the components on would be called a Printed Circuit Assembly or PCA. In the past, track faults may have had to be repaired with hard wires soldered from one part to another. These days this can be done with a component called a **'Jumper'** which is a link component that can be soldered to the track.

There are two types of Printed Circuit Board fixings; **a)** Through Hole and **b)** Surface Mount.

Through-Hole Mounting:

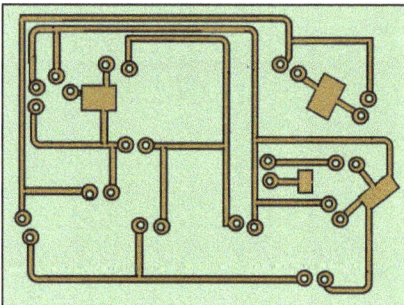

I have just drawn a small **'Through Hole'** printed circuit just to show what one may look like. So this is the copper, solder or 'track' side and the electrodes from the various diodes, capacitors etc. would protrude through the round sections where they would be soldered. As mentioned above these would be a manufacturer specialised circuit for their product not one you would usually set about building yourself.

Surface Mount:

Another type of Printed Circuit Board is what is called the **'Surface Mount'** where components would be mounted on the surface of the board rather than have electrodes that poke through holes and are soldered on the rear. Printed Circuit Boards (PCBs) are made of a rigid material that is a very good insulator/dielectric and can stand quite a large amount of heat. They can be made more flexible for other duties.

These boards come in three classes:

Class 1 PCBs: These circuit boards usually have a life that is fairly reliable, but are not too much of a problem if it fails. For example, on a child's toy.

Apart from the child being upset there would be no danger to human life. Manufacturers tend to keep away from this one for their reputation.

Class 2 PCBs: These are reliable circuits and built into devices where reliability is expected such as televisions, microwaves, computers etc. Again, failure would cause upset but no danger to human life.

Better reliability than Class 1 and manufacturers tend to go for this for their reputation.

Class 3 PCBs: Now these boards are made to have extreme reliability. These are installed into equipment where failure could result in loss of human life, for example in a jet airliner, medical equipment in an operating theatre etc.

Changing Components:

This book explains how the components work and how to test them, however to change components on a printed circuit board, especially Atex Certified Equipment, would require **MANUFACTURER APPROVAL** and with Certified Equipment that would be very unlikely.

Changing components would be considered to be an 'Unauthorised Modification' and any documentation such as a Certificate of Conformity could be rendered null and void. Using the information however may put you on the right track, excuse the pun, to finding the fault. I will say at this stage that 'Through Hole' mounted components would be more difficult to change than Surface Mount.

Faults:

5 x Common Faults on Printed Circuit Boards (PCBs) might include:

1) Black emanating from a particular component which has exploded or overheated.

2) A rust patch on the board where damp or water has attacked it causing discolouration.

3) Circuit boards get smothered in dust and overheat noticed by discolouration.

4) Sometimes components just give up with age. These faults may be very hard to find.

5) Conducting dust such as Carbon has found its way onto the board causing short circuits.

HAVE A GOOD LOOK FIRST. Sometimes the fault may be obvious either by eye or under a magnifying glass. Other not so obvious faults can be found with a multi-meter set for continuity on Ohms (Ω). You may have to remove components to obtain a true test. Broken track can be repaired by soldering a 'wire' link from one end to the other. You must ensure that the multi-meter voltage does not overload the component.

Testing of Printed Circuit Board Components:

Testing of individual components most of the time can usually only be carried out with the component removed from the circuit which, as I have explained above, may invalidate any documentation.

Any modifications to or changes of components on **ANY** Atex equipment could invalidate certification unless Manufacturers are involved and approve changes **IN WRITING**.

Double Sided Printed Circuits:

Most Printed Circuits that I have come across have all of the components on one side of the board and therefore may be determined as having a top and bottom or front and back. However it is possible to come across Printed Circuit Boards which are double sided meaning they have components on both sides of the board.

Multi Layered Printed Circuit Boards:

Substrate	**Made from Fibreglass**
Copper Layer	**Laminated to board with heat**
Solder Mask	**Protects the Track**
Silkscreen Layer	**Helps Label components on the board**

Above is a simple multi layered board, which is more or less a 'Standard' Printed Circuit Board, but boards can be obtained with many more layers and many 'Substrate' etc. This all depends upon the duty that the board is designed to do. **Gold** can also be used in the manufacture of the board.

What are the layers for?

Looking at the diagram above there is the first layer called a **'Substrate';** this is an insulating layer with the **'Copper'** layer laminated to it. A procedure called **'Chemical Etching'** then removes the excess Copper leaving the desired **'tracks'**, which carry the current to the components.

A layer called a **'Solder Mask'** is then applied to protect the copper tracks from corroding and also cuts down the chances of short circuits. This coating would be **green** which gives the actual Printed Circuit Board its green colour.

The Silkscreen layer provides ink that will identify component positions on the PCB. Called a **'Silkscreen'** because it is actually a silk fabric that is stretched over the PCB. All of the information that you see on the PCB such as component names, numbers, polarities, warning information etc. will have been printed onto the board using the Silkscreen.

Vero board

Starting off making your circuit:

Vero Board (below) is very different from a printed circuit and is more what would be used in an electronics hobby. Obviously the components are on the reverse side. With Vero Board you are using the 'Through Hole' method of installing. **Draw a Schematic first!**

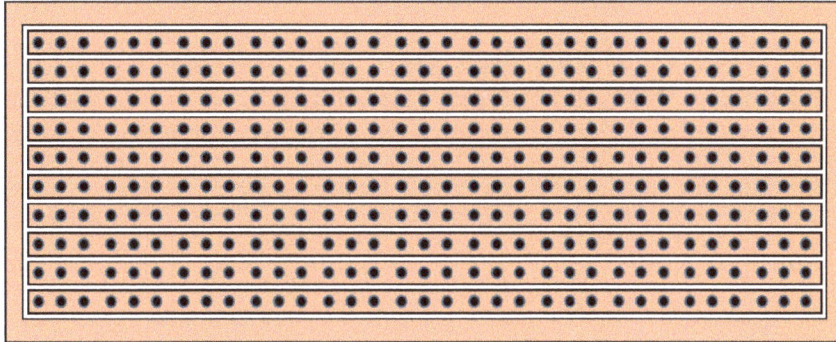

Planning the Components:

We now cut a section of Vero Board large enough to take the components of our circuit. Insert the **'Through Hole'** components onto the top of the board just to ensure that you have the right size and to decide where the strip has to be cut on the underside to ensure that the circuit does not short out and there is plenty of room. Remember at this point you have not soldered anything so there is time to change.

Cutting the Vero Board Copper Strips:

Remove the components and turn the board over to reveal the copper strips. This is where the board strips have to be cut to separate the circuit components.

Underside

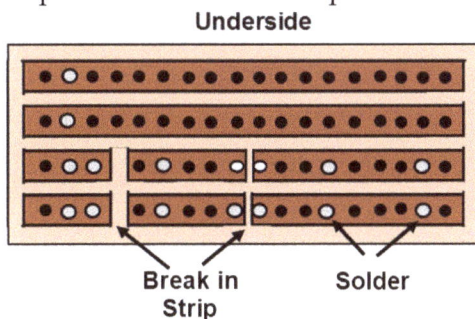

Break in Strip **Solder**

The underside of the board is copper strips as in the diagram to the left, and above this is where the 'through hole' components are soldered into place. The copper strips, if required, must be broken in parts to allow the components to carry out their task, such as a transformer, otherwise the copper strip would short them out. The strip can be carefully drilled to open circuit it or cut with a hacksaw.

Marking the Board:

Vero Board can be obtained with a large number of copper strips for complex circuits. Circuits must be planned very carefully to ensure that all components including transformers of whatever type can be fitted onto the board comfortably.

The way to plan is to mark the stripboard a bit like a chess board with letters A, B, C etc. down the side i.e. a letter on each copper strip, then numbers 1, 2 3 etc. across the top which would in fact be rows of holes. Then bend the component electrodes comfortably and see how they fit. There may be occasions when copper strips have to be physically shorted out with wire or cut/drilled as part of the circuit.

The Final Job:

Finally insert your components and solder them one by one. Always check that each component is inserted into the correct holes in your Vero Board **BEFORE YOU SOLDER!**

Breadboard

Electronic Breadboard, below, looks a little like a fancy crib board and allows you to test individual electronic **'through hole'** components without the awkwardness of trying to hold everything in your fingers i.e. the test probes and the item under test.

You can also build electronic circuits with this board without the permanent task of fixing by solder. I must stress that this is meant to be a **temporary** testing/circuit compiling board.

The Breadboards come in all sorts of shapes and sizes, many with other components already fitted like batteries and switches. The one in the diagram below is equipped with letters along the top and numbers down the side on the main section so you know on which line you are fitting your component.

Inside, under the plastic front, are metal clips which hold your component whilst it is being tested or components if you are building a small circuit. Wire is used to join the components instead of copper strip and the wire is pushed through the hole into the clip just the same as the components.

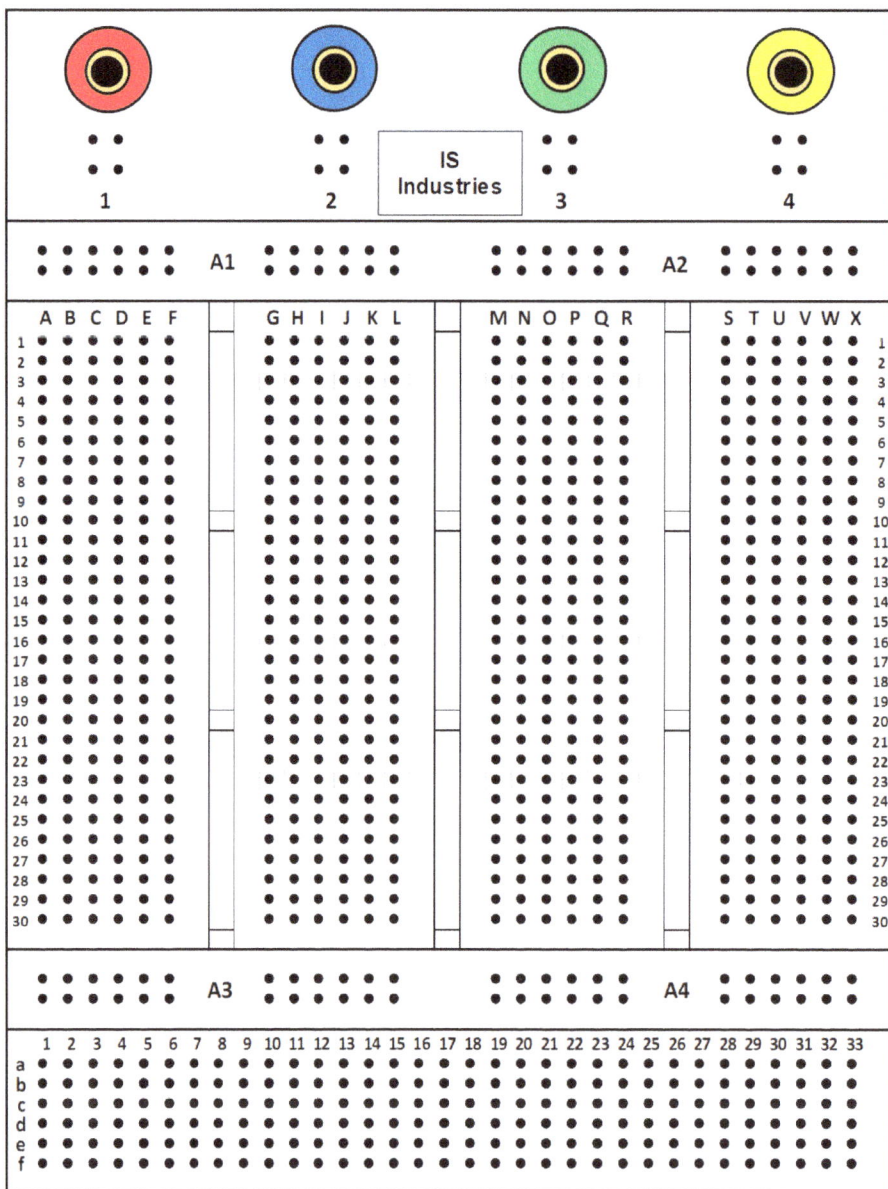

IS Industries

A1   A2

A B C D E F   G H I J K L   M N O P Q R   S T U V W X

A3   A4

Printed Circuit Board Questions and Answers:

Q1 – Why are Printed Circuit Boards predominantly green?

A1 – **A COATING.** When the tracks are complete the board is totally covered with a solder mask to stop the tracks from oxidising in the air and this coating gives the PCB a green colour.

Q2 – What is 'Silkscreen'?

A2 – **FOR LOCATING COMPONENTS.** The location is printed onto the Silkscreen making it easier when first making up the board.

Q3 – Can PCBs be flexible?

A3 – **YES.** Sometimes they are required to be flexible to fit the equipment.

Q4 – On a PCB what is the difference between Clearance Distance and Creepage Distance?

A4 – **a) CLEARANCE DISTANCE.** The physical distance in air between two components.

  b) CREEPAGE DISTANCE. The distance between two conducting parts i.e. Track.

Q5 – What is Gerber Format?

A5 – **A DESIGN FORMAT.** Used to design Printed Circuit Boards. Similar to CAD (**C**omputer **A**ided **D**esign).

Q6 – Can I re-use an old PCB for another duty?

A6 – **YES.** If the track routes to where you want it to go and voltage & current parameters match. You may want to look at Vero Board if you are building your own PCB.

Q7 – Can components be mounted on both sides of the PCB?

A7 – **SOMETIMES.** If it is designed that way.

Q8 – What is a PCB Breadboard?

A8 – **A NON SOLDER DESIGN BOARD.** Where components can be easily mounted and changed around in the circuit design stage. Also a stable mount for testing.

Q9 – What is a Motherboard?

A9 – **A MAIN BOARD.** Inserted in appliances by the manufacturer, i.e. in a computer.

Q10 – Are there other Manufacturer Boards besides the Motherboard?

A10 – **YES.** There are Daughter Boards which are connected to the Motherboard. Seems a bit sexist!

Q11 – Can components be removed once they are soldered in?

A11 – **YES.** But sometimes that is more difficult than it seems at first. A solder sucker will make it easier. Surface mounted components are easier to remove.

Q12 – On the printed circuit board there are large oblong areas as well as tracks. What are they?

A12 – These are called **'PLANES'** and have tracks leading to them.

Q13 – Can I replace & repair Printed Circuit Boards on Atex Certified Equipment?

A13 – **NO.** Not without Manufacturer approval **IN WRITING.**

The Printed Circuit Fuse Section:

Fuses in printed circuits can take many forms, but the one thing they have to be is **FAST.** They have to blow at the very first sign of surge or overload. Time can cause catastrophic failure of electronic components or fire.

The fuses can be separate or as below built into the Printed Circuit Board as a weak link. This type, of course, cannot be repaired easily by a technician. Sometimes they can be bought as a component and soldered into the board.

What has got to be looked into, regardless of the fuse type, is **what caused it to blow in the first place? Replaceable type fuses must not be renewed until the failure/cause of the fuse blow is investigated.**

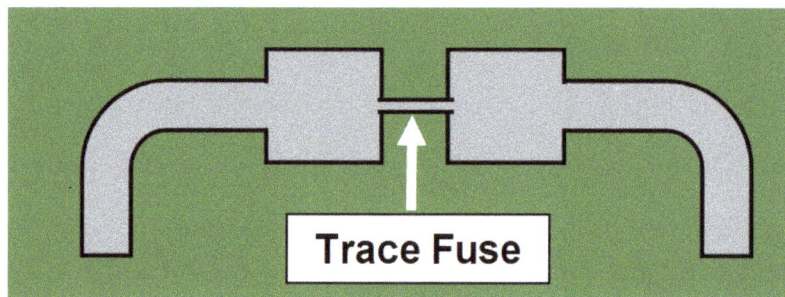

Trace Fuse

Obviously 'Trace' fuses such as the one in the diagram above would be extremely difficult for a Technician to replace, but at least they will know the problem, but not the cause without further investigation. They come in a range of different shapes some like the diagram above others more complex! **Think twice about soldering some fuse wire across the gap.** They tend to burn the PCB when they blow. They maybe can be obtained as a component.

In this section we look at several different types of fuse and how we might investigate why they may have blown. Replacing electronic protection fuses with any other types can, as stated above, cause catastrophic failure of components or even fires.

In the case of power circuits, sometimes replacing the fuse can blow the fault clear, but in the case of electronics, integrated circuits or intrinsically safe equipment that is very rarely the case.

Please note that this book is only meant to guide you into finding faults on electronic printed circuits so that you can endeavour to be sure as to what the problem is, so you know which part needs to be changed.

Conventional fuses such as glass or any cartridge-type fuse can be completed by the technician after a very thorough investigation as to the cause of the fuse blowing.

There are **'eFuses'** these days which are an electronic component which uses a Field Effect Transistor (FET) to monitor the supply with a sensing resistor. Some are resettable. They are extremely fast. Manufacturers may give advice on upgrades.

NOTE:

REPAIRS AND REPLACEMENTS TO PRINTED CIRCUITS ON ATEX CERTIFIED EQUIPMENT FOR HAZARDOUS AREAS, SUCH AS ZONES, CAN ONLY BE DONE WITH MANUFACTURER APPROVAL IN WRITING.

FAILURE TO CONFORM TO THIS MAY BE CLASSED AS AN UNAUTHORISED MODIFICATION AND MAY RENDER ANY CERTIFICATION DOCUMENTATION NULL AND VOID.

PCB Board Fuses:

Conventional Fuses:

Conventional fuses have characteristics such as Fusing Factor, Discrimination etc., but would these types of fuses, be they cartridge, rewire able etc., be fast enough to protect electronic and integrated circuits? So how does a fuse blow? Well firstly the word 'Fuse' means to melt, so the fuse melts when the current reaches a certain value.

Let us take a domestic 5Amp Fuse, it does not blow when the current reaches 5.1Amps otherwise motors in lawnmowers, vacuum cleaners, washing machines etc. would never get started because of their 'start' current. The fuse may heat up but not blow unless the motor did not turn when it was meant to, or took too long to spin up.

Electronic Protection Fuses:

With electronics the above is not true as there are many very delicate components such as integrated circuits etc., that require an **ULTRA** fast blow fuse to operate at the first sign of trouble.

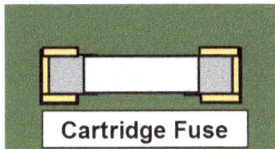

Cartridge Fuse

Sometimes the Printed Circuit Board has a good old-fashioned Cartridge Glass Fuse in place. This fuse will be a fast blow PCB Protection fuse and should not be replaced with anything other than the manufacturer's fuse. The good thing is that they are replaceable. Why did it blow to start with?

Printed circuit Fuses come in many 'through hole' shapes and sizes and look like many other components on the board. Sometimes the board is marked as what components are which makes our fault finding life a little easier. Do not replace any fuses with different types if they blow. You may not be allowed to replace on Certified Equipment!

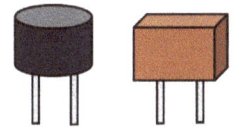

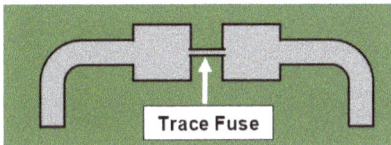

Trace Fuse

As per the diagram on the left, manufacturers will actually build the fuse into the printed circuit trace on the board. The fuse may have an identification mark of 'Fuse' in letters on the board. May be bought as a component – seek advice. **Why did it blow?**

Through Hole Fuses for printed circuit boards may look exactly like a Resistor or Diode. They are usually dark in appearance and actually soldered, as other components, onto the board.

Fault Finding:

I have put the question a couple of times above; **WHY DID IT BLOW?** Was it sustained overcurrent, surge etc.? Whatever caused the blow must be investigated before the circuit is replaced, and **must be like for like!**

Take a magnifying glass and very carefully look at the whole of the board together with each component. You are looking for common things like:

1) Discolouration of the board, which might be caused by water, damp or heat.

2) Discolouration of components due to heat etc.

3) Black emanating from anywhere around any components.

4) Any components completely obliterated i.e. by a small explosion from within.

5) Any conducting contaminates got onto the board through any seals such as carbon?

6) Obvious short circuits on the printed circuit board track caused by debris from something else.

7) Any electromagnetic issues such as interference?

I keep saying this but it must be followed that if the equipment is Atex Certified, unless the Fuse Unit is a simple Glass Cartridge Fuse that can be changed easily, then manufacturers must be consulted.

PCB Fuses Questions and Answers:

Q1 – Can PCB Glass Fuses be obtained in different currents?

A1 – **MOST DEFINATELY.** Look at the metal cap for the current.

Q2 – Are there AC & DC Fuses on Printed Circuit Boards?

A2 – **YES.** They are different. AC Fuses usually deal with larger currents.

Q3 – Can I change a soldered Fuse on a PCB?

A3 – **NOT WITHOUT MANUFACTURER APPROVAL.** If the equipment is Atex Certified.

Q4 – Does a Fuse have a voltage rating?

A4 – **YES.** Usually before the letters AC or DC.

Q5 – Can Fuses be fast blow?

A5 – **MOST DEFINATELY.** Fast Blow Fuses may have the letter 'F' on the cap and may be a Glass Fuse with a very thin element.

Q6 – Do electronics require Fast Blow Fuses?

A6 – **MOST DEFINATELY.** Electronics would require an **ULTRA** Fast Blow Fuse! Look for 'F' or 'FF' on the cap. Element usually very thin. (No springs.) Remember that in this case you are not only protecting components such as Diodes & Transistors, but also there may be Integrated Circuits involved. **There are 'eFuses' these days which are very fast.**

Q7 – Can I replace a trace Fuse on my printed circuit board with another type?

A7 – **NOT WITHOUT MANUFACTURER APPROVAL.** If the equipment is Atex Certified. In other cases, check that the fuse characteristics and size match. Ask manufacturer for upgrade.

Q8 – What if my Glass Fuse has a spring near one cap on the element inside of the glass?

A8 – **USUALLY SLOW BLOW.** Look for an 'S' or a 'T' on the metal cap. Not for electronics.

Q9 – If I cannot find the correct Fuse can I use a normal Domestic Fuse on my PCB?

A9 – **DEFINATELY NOT.** Domestic fuses would be too slow. This is not just a case of getting the circuit to work. The previous fuse blew for a reason. Putting slow speed fuses in may compound the problem or even cause a fire!

Q10 – Are PCB Fuses surface or 'through hole' mounting?

A10 – **BOTH.** Depends upon the PCB design. Can also be trace (track) fuses.

Q11 – Can fuses just blow without a fault?

A11 – **YES.** Mainly if the fuses are old and have been working on and off on high amperage devices such as starting a motor regularly or switching a high number of lights. It would be unusual for PCB Fuses to just blow as they should not be under much load at all. Local lightning strikes can cause electronic fuses to blow.

Always check a PCB thoroughly before replacing fuses as you could cause a spike or surge if the fault has not been rectified.

The Printed Circuit Switch Section:

We tend to take switches for granted in electronics, but there are lots of different types of switches which are used for different purposes.

Switches on Printed Circuit Boards come in many shapes, sizes and duties. They can be push button or toggle switches, for example, which simply switch the circuit on and off or DIP switches like the diagram below left, which are used to program certain equipment to function in a certain way.

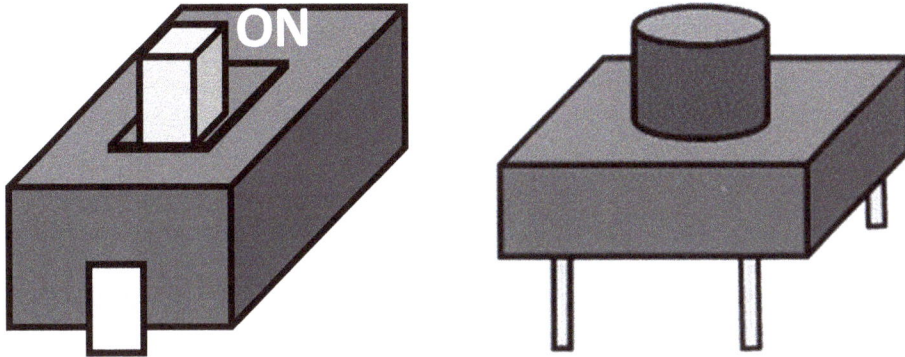

Some switches can be 'through hole' mount or others 'surface mount' depending upon the design. The switch can even be remote so instead of it being part of the PCB it is actually mounted on the item of equipment.

DIP switches are not usually single switches as in the diagram above, they are usually multi gang switches and you will see them on special relays such as high voltage overload relays. They actually program the relay to do certain built-in tasks by the manufacturer.

If you design an electronic circuit always include a switch and, as mentioned in an earlier section, a fuse so that if anything goes wrong it can be isolated.

When people think of a switch they assume that it is just ON & OFF, but that is not necessarily so as some switches have multi functions such as two different ON positions with an OFF position in the middle so the switch would, in fact, be ON – OFF – ON.

Let us have a look at a range of switches and how they may fit into a Printed Circuit Board.

NOTE:

ON ALL ATEX EQUIPMENT FOR HAZARDOUS AREAS SUCH AS ZONES, REMOVAL/CHANGING OF SWITCHES FROM A MANUFACTURER'S PRINTED CIRCUIT BOARD CAN ONLY BE CARRIED OUT WITH MANUFACTURER APPROVAL IN WRITING.

FAILURE TO CONFORM TO THIS MAY BE CLASSED AS AN UNAUTHORISED MODIFICATION AND MAY RENDER ANY CERTIFICATION DOCUMENTATION NULL AND VOID.

ALSO THE DESIGN, MANUFACTURER, TYPE, ETC., OF THE SWITCH COMPONENTS MUST NOT BE CHANGED IN ANY WAY.

Printed Circuit Board Switches:

General Description:

PCB switches come in all shapes, types and sizes, some of which I have shown below but there will be many more types. Designing a switch onto a PCB obviously suggests that certain circuits require switching on or off.

When mounting the switch sometimes it is better to mount onto a back plate and then onto the board rather than just via a 'through hole' mount, especially if the switch is used regularly, but switches can be mounted directly and soldered on a 'through hole' fixing.

Obviously if a switch has to be fitted then that suggests that a Printed Circuit requires switching on and off to make it work.

PCB Pushbutton Switches:

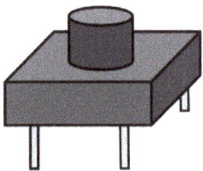

PCB Pushbutton Switches are mounted directly onto the printed circuit and are used as you might think as normal switches in miniature to switch particular electronic circuits on and off. They do come as a through hole fitted unit that is soldered onto the board, usually single pole activation. The switch mechanism itself can be round as in the diagram on the left, or square.

PCB DIP Switches:

PCB **DIP** Switches (**D**ual **In**-line **P**ackage) can be single, as in the diagram on the right, or multi-switches. These are usually used to set a particular sequence for a particular set of functions to happen. They are not usually operated on a single switch basis to switch a device on and off, for example, HV Electronic Protection Relay functions.

PCB Toggle Switches:

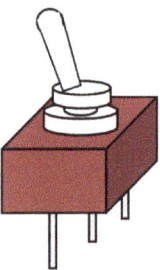

PCB Toggle Switches are mounted directly onto the printed circuit and are used, as you might think, as normal switches in miniature to switch particular electronic circuits on and off. They do come as a through hole fitted unit that is soldered onto the board.

These switches in some cases have just two positions ON – OFF whilst others have three different positions so in fact they would be an ON – OFF – ON switch. Their toggles may be conventional as in the diagram on the left, or flat bat shaped.

PCB Slide Switches:

PCB Slide Switches are mounted directly onto the printed circuit and are used, as you might think, as normal switches in miniature to switch particular electronic circuits on and off. They do come as a through hole fitted unit that is soldered onto the board, usually single pole activation. The switch mechanism itself can be round as in the diagram on the right showing the slider. Larger than a PCB DIP Switch.

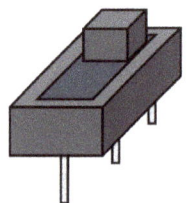

PCB Tactile Switches:

Many PCB Switches are called **'Tactile'** Switches simply because they require human touch to activate them! Examples of 'Tactile' Switches would be the PCB Pushbutton Switch, PCB Toggle Switch and PCB Slide Switch.

PCB Switch Questions and Answers:

Q1 – What are DIP Switches?

A1 – **DUAL INLINE PACKAGE.** They are usually not an individual switch, they are usually in multi-gang situations that are used to program a piece of equipment for a number of functions for it to perform.

Q2 – Are DIP switches toggle switches?

A2 – **NO.** They can be circular, slide or rocker.

Q3 – Is the switch mounted directly onto the Printed Circuit Board?

A3 – **YES AND NO.** Many are direct 'Through Hole' mounting. Some switches, however, can be mounted on a plate before being mounted on the board. Some may be remote on equipment.

Q4 – Are PCB switches just two functions **ON** or **OFF**?

A4 – **NOT NECESSARILY.** Depending upon their duty they can be three functions

i.e. **ON – OFF – ON**.

Q5 – Why are many PCB switches called 'Tactile' switches?

A5 – **HUMAN TOUCH.** Is required to operate them.

Q6 – Are PCB switches single or double pole?

A6 – **SINGLE POLE.** Usually when PCBs are involved.

Q7 – What is a MEC Switch?

A7 – **DANISH MINIATURE PCB SWITCHES.**

Q8 – Can the switches be remote from the PCB?

A8 – **MOST CERTAINLY.** And many are actual switches on the side of the equipment.

Q9 – What is the difference between a PCB switch and a normal switch?

A9 – **LOWER RESISTANCE.**

Q10 – Can I obtain a PCB switch with a built in indicator light?

A10 – **YES.** Usually LED these days.

Q11 – Can I change the switch for another type?

A11 – **NOT IF IT IS 'ATEX' EQUIPMENT.** Without manufacturer approval in writing.

Q12 – Are PCB switches Ingress Protection (IP) rated against the weather?

A12 – **STRANGELY ENOUGH.** Many of them are.

Q13 – Can PCB switches include gold in their manufacture?

A13 – **YES.** And the Printed Circuit Board itself can contain gold plating in its manufacture. Of course anything that includes gold will be very expensive.

The Printed Circuit Transformer Section:

The Transformer used on a Printed Circuit will usually be a 'Step Down' Transformer meaning that the voltage coming into the device primary winding(s) will be stepped down to a lower voltage by the secondary winding(s).

It is possible to obtain 'Step Up' Transformers where the voltage coming into the primary winding(s) is 'Stepped Up' by the secondary winding(s). It is all to do with winding ratio.

It is not unusual for PCB Transformers to have two secondary windings as in the diagram below and also not that unusual for it to have two primary windings depending upon the design of the circuit.

PCB Transformers can be 'centre tapped' if, for instance, they are being used on a small Electronic Inverter which uses Transistors such as the project that I have included at the end.

Just be a bit careful with the use of Auto Transformers, which, of course, just have one winding. Auto Transformers have been known to short out and the **FULL** output is the result.

When we talk about an Auto Transformer just think of a one bar electric fire. The element Neutral would be one connection and the voltage would depend upon how far along the coil towards the live I touched for the second connection and voltage.

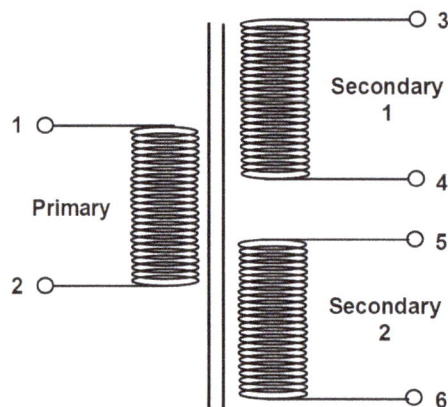

We will discuss what criteria you would require when deciding what size and type of Transformer to choose when designing your circuit.

Let us have a look at how these transformers might be designed into our printed circuit and how many types of PCB Transformers there are.

NOTE:

ON ALL ATEX EQUIPMENT FOR HAZARDOUS AREAS SUCH AS ZONES, **REMOVAL/CHANGING OF TRANSFORMERS FROM A MANUFACTURERS' PRINTED CIRCUIT BOARD CAN ONLY BE CARRIED OUT WITH MANUFACTURER APPROVAL IN WRITING.**

FAILURE TO CONFORM TO THIS MAY BE CLASSED AS AN UNAUTHORISED MODIFICATION AND MAY RENDER ANY CERTIFICATION DOCUMENTATION NULL AND VOID.

ALSO THE DESIGN, MANUFACTURER, TYPE ETC., OF THE TRANSFORMER COMPONENTS MUST NOT BE CHANGED IN ANY WAY.

PCB Transformer Questions and Answers:

Q1 – Will a Transformer work on DC?

A1 – NO. It requires induction which it can only get from AC.

Q2 – What are the coils called in a Transformer?

A2 – INPUT. Primary, OUTPUT. Secondary.

Q3 – Can a Transformer step the voltage up as well as down.

A3 – YES. But it would have to be designed correctly as per its turns.

Q4 – Can a PCB Transformer have two secondaries?

A4 – YES. and sometimes two primaries.

Q5 – What is a Through Hole Transformer?

A5 – The transformer has ELECTRODES that fit through the holes in a PCB ready to be soldered.

Q6 – Can Auto Transformers be used?

A7 – YES. But be careful; they have been known to short and give full voltage. Be careful using them where, if the voltage suddenly went to maximum, it may cause the circuit harm or a fire.

Q8 – What is the difference between a PCB Standard Transformer and a Medical one?

A8 – HIGHER PROTECTION on the Medical Transformer as it may feed Medical Test Equipment and Patients and Staff must be protected.

Q9 – What dissipates the Transformer Heat?

A9 – Usually the LAMINATED CORE. Transformer windings also tend to be open to air flow.

Q10 – What is a Balun Transformer?

A10 – A BALANCED TRANSFORMER.

Q11 – Can Transformers be obtained that mount on the surface and not Through Hole?

A11 – MOST CERTAINLY. Heat would be the biggest enemy here, weakening the solder mount.

Q12 – Can I change PCB transformers for another type?

A12 – DEFINITELY NOT on Atex equipment without manufacturer approval in writing. It is unwise to completely change any component in any equipment without professional design advice.

Q13 – Can a Transformer Primary Windings melt into the Secondary and produce Primary Voltage to the circuit?

A13 – NOT USUALLY. In most transformers there is an earthed core for the melting Primary Voltage to go through before it gets to the Secondary Windings so fuses would blow.

Q14 – What is an 'Isolation' Transformer?

A14 – THE SECONDARY winding is not referenced to earth.

The Printed Circuit Relay Section:

Relays these days can be conventional Electro-mechanical relays with a solenoid and moving contacts (EMR) or in fact 'Solid State' (SSR) relays with no moving parts.

These PCB Relays can be obtained with electrodes ready for 'through hole' fitting and can be soldered directly onto the printed circuit board.

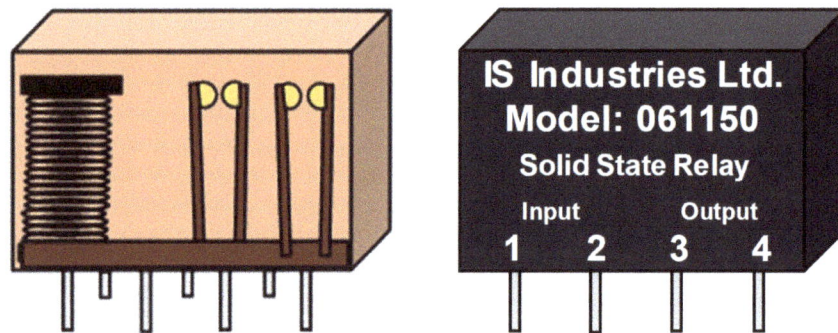

Solenoid Relays come with a range of internal components as below:

1) Solenoid coil with an iron core – Iron is used so that any magnetic field will not cause permanent magnetism.

2) Armature – the section that is attracted by the solenoid causing the contacts to open & close.

3) The Yoke – The metal part of the solenoid holding everything in place.

4) Fixed Contacts – as the name suggests, these are fixed in place and never move.

5) Moving Contacts – as the name suggests are connected to the armature and open and close on a hinge mechanism against the fixed contacts.

6) Normally Open Contacts (N/O) – coil de-energised these contacts are open.

7) Normally Closed Contacts (N/C) – coil de-energised these contacts are closed.

Solid State Relays (SSR): These Solid State relays often convert an electronic signal into an optical signal (light) and a device called an Optical Isolator then collects the light signal and, via an internal device called a Photosensitive Transistor, the light is converted back to an electronic signal by the Transistor to control other components, so no moving parts at all. These may be called 'Galvanic' in their design.

Interposing Relays: Relays may also be what are called Interposing Relays; this is where the Relay Coil may be at quite a low voltage, let us say 24 Volts, yet the contacts are switching a much higher voltage, let us say 110 Volts or higher.

NOTE:

ON ALL ATEX EQUIPMENT FOR HAZARDOUS AREAS SUCH AS ZONES, REMOVAL/CHANGING OF RELAYS FROM A MANUFACTURERS PRINTED CIRCUIT BOARD CAN ONLY BE CARRIED OUT WITH MANUFACTURER APPROVAL IN WRITING.

FAILURE TO CONFORM TO THIS MAY BE CLASSED AS AN UNAUTHORISED MODIFICATION AND MAY RENDER ANY CERTIFICATION DOCUMENTATION NULL AND VOID.

ALSO THE DESIGN, MANUFACTURER, TYPE ETC., OF THE RELAY COMPONENTS MUST NOT BE CHANGED IN ANY WAY.

PCB Relays:

Electromechanical Relays in General:

Printed Circuit Board relays are obviously very small. They come with AC or DC coils and with normally open (N/O) or normally closed (N/C) contacts. Normally open contacts are in the open position when the relay coil has no power onto it and normally closed contacts are in that position with no power onto the coil. Just take a relay as a switch that is operated electronically.

Difference between a Relay and a Contactor:

There is not a great deal of difference between a Relay and a Contactor as they both have a solenoid that operates either normally open (N/O) or normally closed (N/C) contacts.

One of the main differences is the loads that they switch. A **Contactor** will be much larger than a relay and may be used, for instance, in an electric motor starter to start and stop an electric motor.

As you could appreciate the load on the contacts when starting the motor would be quite high. Also the main contacts in this unit would be normally open (N/O), closing to start the motor. Auxiliary contacts can be N/O or N/C.

A **Relay** on the other hand only switches very low currents and can be manufactured so small that it can be fitted onto a printed circuit board. The contacts in this case can be N/O or N/C.

Solid State Relays:

Sometimes these PCB Relays can be solid state (SSR)! Many relays used today and in the past are what is called standard electromechanical (EMR). This is where the relay is a solenoid where there is a coil that receives a signal, as above, AC or DC depending on the type, and this physically opens or closes a set of contacts.

A PCB Solid State Relay (SSR) has no moving parts. Again as a standard EMR the SSR receives an AC or DC signal, depending upon its specification, which it converts to light to carry out the switching which can be done using a light emitting diode for the light signal, and collector to turn the signal back to electronic, similar to an optical isolator. The action would be galvanic with no actual contacts.

Sum up:

So to sum up: the Printed Circuit Board Relay (EMR) can have a solenoid which receives an AC or DC signal depending upon its design specification, which in turn operates a set of contacts which can be normally open or normally closed and, on receiving a signal, moving parts either open or close those contacts. The relay may be a bit more complex than my illustration on the left. They come with through hole fixing if required, ready to be soldered onto the PCB.

A Printed Circuit Board Solid State Relay (SSR) design may look similar to the diagram on the right and, if required, come with through hole fixing ready to be soldered directly onto the printed circuit board. The relay configuration may be marked on the side and may be a bit more complex than in my diagram.

IS Industries Ltd.
Model: 061150
Solid State Relay

Input		Output	
1	2	3	4

Schematic Diagrams:

Electrical or Electronic Schematic Diagrams are always drawn with circuits and components in the de-energised position. So for instance if relay contacts are supposed to be normally open (N/O) at rest with no power, then that is how they will appear on the drawing.

Types of PCB Relay:

1) **SPDT 1 (With Mechanical Switch):** Single Pole Double Throw: Single Pole relays operate one circuit only. Throw means the number of contacts N/O or N/C, and this relay has two. **Does not latch. Uses a metal switch for operation.**

2) **SPDT 2 (With Magnetic Contacts):** Single Pole Double Throw: Single Pole relays operate one circuit only. Throw means the number of contacts either N/O or N/C, and this relay has two. **Does not latch. Uses a magnetic core for operation.**

3) **SPST:** Single Pole Single Throw: Single Pole relays operate one circuit only. Throw means the number of contacts N/O or N/C, and this relay has one. **This relay latches.**

4) **DPST**: Double Pole Single Throw: Double Pole relays operate two circuits. Throw means the number of contacts N/O or N/C, and this relay has one.

5) **DPDT:** Double Pole Double Throw: Double Pole relays operate two circuits. Throw means the number of contacts N/O or N/C, and this relay has two.

6) **Solid State:** Relays are used these days because of their faster switching capabilities without being subject to damage.

Switch Speed Protection:

Many standard PCB relays will have a built in speed protection which will stop them being switched too many times at speed which could damage the relay and equipment.

Thermal Protection:

These Thermal Relays will trip if the temperature goes over a pre-set value.

Testing a Solenoid Relay:

Remove the relay from the circuit as it is usually difficult to test while connected in.

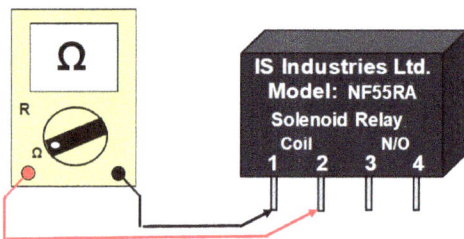

1) Find the two terminals on the relay that are the coil.

2) Turn the selector switch on the multi-meter to ohms (Ω)

3) Measure the resistance of the coil.

The multi-meter should show a value in ohms, but unless you know the true manufacturer's coil resistance you can only assume it is correct with no coil shorts.

If there is a coil resistance that seems to be around the correct resistance you may want to test its full operation. The contact in my diagram above is normally open (N/O) so energising the relay should cause it to close. Connect as the diagram below, ensuring that the Voltage wiring cannot come into contact with the multi-meter probes. This may be done by using connectors.

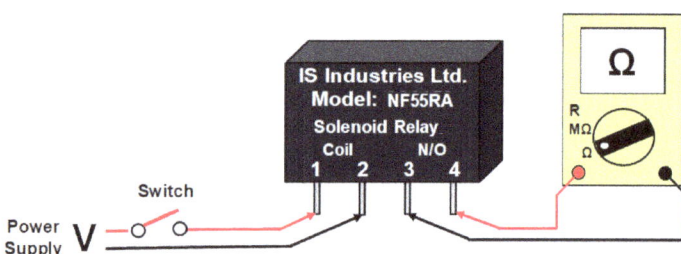

1) Ensure your voltage is correct for the relay coil AC – DC and the size of the voltage.

2) As above do not let the voltage come into contact with the resistance probes.

3) Insert a switch to switch on and off!

PCB Relay Questions and Answers:

Q1 – If a Relay is Single Pole what does that mean?

A1 – ONE circuit is operated by the Relay.

Q2 – If a relay is Double or Single Throw what does that mean?

A2 – PAIRS of contacts. Single throw has one pair of contacts.

Q3 – When contacts are said to be normally open or normally closed what does that mean?

A3 – N/O OR N/C is the state of the contacts at rest with no power on the Relay.

Q4 – What is the difference between a Relay and a Contactor?

A4 – LOAD. A Contactor will switch higher loads and its main contacts are usually N/O.

Q5 – Can a Relay be Solid State?

A5 – MOST CERTAINLY these days with no moving parts. They would be SSR.

Q6 – What does EMR mean?

A6 – ELECTROMECHANICAL. It has moving parts.

Q7 – Has a PCB Electromechanical relay got switching speed limitation?

A7 – YES. They will have in order to stop Relay and Circuit damage.

Q8 – Do solid state relays have speed limitation.

A8 – NO. They are much faster than Electromagnetic Relays.

Q9 – Can I use a multi-meter set on continuity to test the Relay Coil?

A9 – YES. But unless you know what a good coil resistance is supposed to be you will not know if there are any shorts on the coil wires.

Q10 – Can I test a Solid State Relay (SSR) the same as an Electromechanical Relay (EMR)?

A10 – YES. Ensure that voltages are correct. There is no coil to test.

Q11 – Can Normally Open (N/O) contacts be changed to Normally Closed (N/C)?

A11 – NO. In most cases you will either have to obtain a Relay with Normally Closed Contacts or a Relay with a Double Action Contact.

Q12 – Will the PCB Relay trip if the temperature gets too high?

A12 – YES. Most PCB Relays will have this feature. Check manufacturer's data.

Q13 – Will the PCB Relay protect against overcurrent?

A13 – YES. Most PCB Relays will have this feature. Check manufacturer's data.

Q14 – Will the PCB Relay protect against high voltage spikes?

A14 – YES. Most PCB Relays have this feature. Most will also protect against an electrostatic spike (ESD).

The PN Junction Section:

This section discusses how a PN Junction works, and where it fits into various printed circuit components.

Understanding a PN Junction can be the key to understanding many electronic components such as Diodes, Transistors, Thyristors, etc.

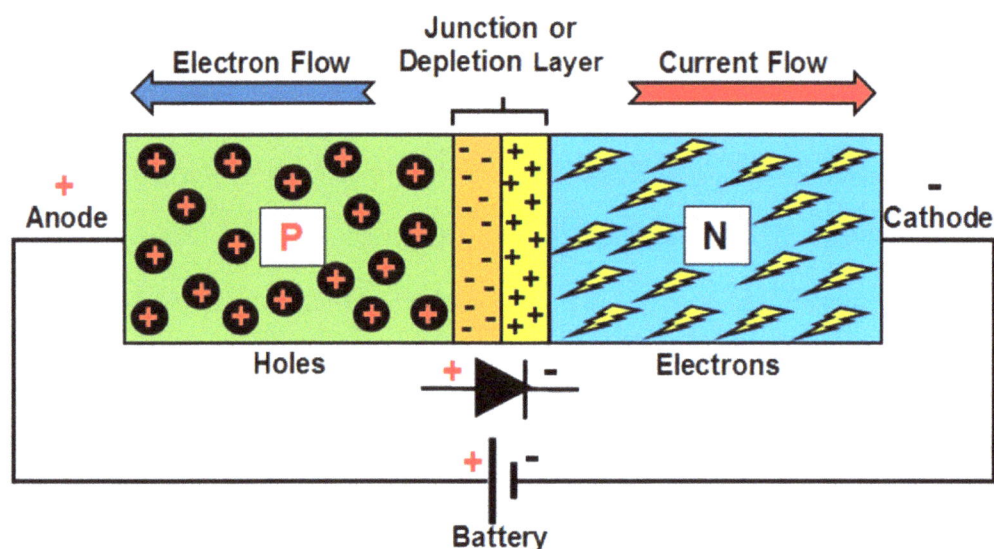

I am sure that most of the time 'I am teaching my grandmother to suck eggs' in my explanations, but I believe that it does no harm to go over the theory.

PN Junctions are in: Diodes, Solar Cells, Transistors, Integrated Circuits, LEDs etc. They can be Forward (Positive) or Reverse (Negative) Bias discussed below.

Positive Bias: Supply Positive (+) connected to the 'P' material and the Supply Negative (-) connected to the 'N' material. Discussed later.

Negative Bias: Supply Negative (-) connected to the 'P' material and the Supply Positive (+) connected to the 'N' material. Discussed later.

The main material here for our PN Junction can be Silicon. The Silicon can be what is called 'doped' to give the PN effect. Doping means that in this case another substance is added to the Silicon to give it either a 'P' property or an 'N' property.

So if we look at the above diagram we will see two Silicon Semiconductor sections, one Green 'P' (Positive) and the other section Blue 'N' (Negative).

We can arrive at these two Semiconductor sections by doping the 'N' Silicon section (Blue) with Antimony and doping the 'P' Silicon section (Green) with Boron. Put these two together and we have our PN Junction.

In the diagram above I have drawn a Diode to give you an idea of how this component would fit into our explanation and how we arrive at the direction of current.

We will discuss these in this section.

The PN Junction:

Outline:

The Diode & Bipolar Transistor is made up of a PN junction so what exactly does this mean? What happens with the junction to enable the Diode or Transistor to operate? Well firstly we need to know what a Diode and a Transistor is.

A Diode is an electronic device, the most well known ones are used to change AC to DC, whilst others are used to, for instance, protect other delicate electronic systems as a safety device and their uses go on. A Transistor is an electronic switch or amplifier.

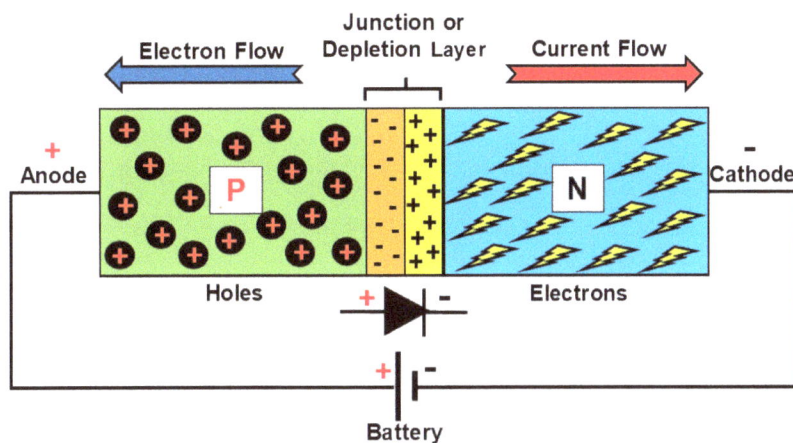

Description:

So what is a PN Junction? Looking at the diagram above we have three different layers. The green layer is a **Germanium or Silicon** semiconductor material doped, meaning that another material has been added, with a positive **'Trivalent'** (valence of 3) material that predominantly causes 'holes' for example.

'Boron' could be the blue material doped with a negative **'Pentavalent'** (valence of 5) material, for example, **'Antimony'**, predominantly containing electrons.

Separating the two Green and Blue regions, where they meet is a yellow interface material which we will call the **'junction'** or **'depletion layer'** which is fused to the other layers by special means.

Forward Bias:

Before we go much further just let me mention **'zero'**, **'forward'** and **'reverse' bias**. The above diagram is connected in **'forward'** bias meaning that the **positive** side of the battery is connected to the **positive** side of the 'P' region.

Reverse Bias:

So **'reverse'** bias, for example, would be when the battery is connected the other way i.e. negative to 'P' side of Diode.

Bias Effect:

So what effect would this have? Well in **'forward'** bias the depletion region would decrease, allowing electrons from the 'N' region to flow across the junction to the holes in the 'P' side. In 'reverse' bias the junction or depletion zone would expand, because the electric fields set up by the voltage would be both in the same direction so no electrons could flow and holes would move away from the junction.

When we talk about the 'reverse' bias we must look at Diodes such as **Avalanche** and **Zener Diodes**. If the voltage does get high enough, even with reverse bias, the depletion zone will break down and allow electron flow. This is called the 'breakdown voltage' and is used very effectively in Avalanche and Zener Diodes to guard other electronic units against surges of high voltage.

Depletion Zone:

When the 'P' is fused into the 'N' of the Diode or Bipolar Transistor, the junction is formed and some free electrons move immediately across from the negative to the positive side and some holes move across from the positive to the negative side to form the **'Depletion Zone'.** This process is called **'diffusion'**.

Once this action is completed with no voltage applied yet, then the junction is depleted of free electrons, which is why it is called a **'Depletion Zone'**. **'Zero'** bias is when there is no voltage applied to the Diode or Transistor.

A voltage must be supplied at a certain level before the Diode is activated and this is when the voltage at the Anode (+) is higher than the voltage at the Cathode (-) and a Gate voltage for the Transistor.

Questions & Answers on PN Junctions:

Q1 – What is the meaning of PN Junction in 'Forward' Bias?

A1 – POLARITY. The battery positive terminal is connected to the Junction anode and the negative terminal connected to the Junction cathode.

Q2 – What is the meaning of PN Junction 'Reverse' Bias?

A2 – POLARITY. The battery positive terminal is connected to the Junction cathode and the negative terminal connected to the Junction anode.

Q3 – What is doping?

A3 – IMPURITIES added to a semiconductor to increase conductivity.

Q4 – What is a PN Junction Depletion Region?

A4 – SEPARATION LAYER between the 'P' Region and the 'N' Region where there are no electrons or holes.

Q5 – What material would be at the anode of the PN Junction?

A5 – DOPED GERMANIUM OR SILICON for example.

Q6 – What material would be at the cathode of the PN Junction?

A6 – DOPED BORON for example.

Q7 – What would be the doping material for the PN Junction cathode?

A7 – PENTAVALENT MATERIAL possibly Antimony.

Q8 – What would be the material at the PN Junction anode?

A8 – TRIVALENT MATERIAL.

Q9 – What would the current be in a reverse bias PN Junction Diode?

A9 – VERY SMALL INDEED!

Q10 – Which part of the PN Junction contains holes and which part contains electrons?

A10 – 'P' CONTAINS HOLES & 'N' CONTAINS ELECTRONS.

The Transistor Section:

In this section we look at all of the different types of Transistor, how they work and how they fit into our electronic world. Technicians look at a Transistor fitted into a Printed Circuit and totally take this component for granted. There are so many different types of Transistor.

The main difference between a Transistor and a Thyristor is that most Thyristors latch when power is applied to the base whereas Transistors do not.

When designing a Printed Circuit or even building electronics as a hobby, where Transistors are involved the most important question you need to ask yourself is:

"What do you require the Transistor to do?"

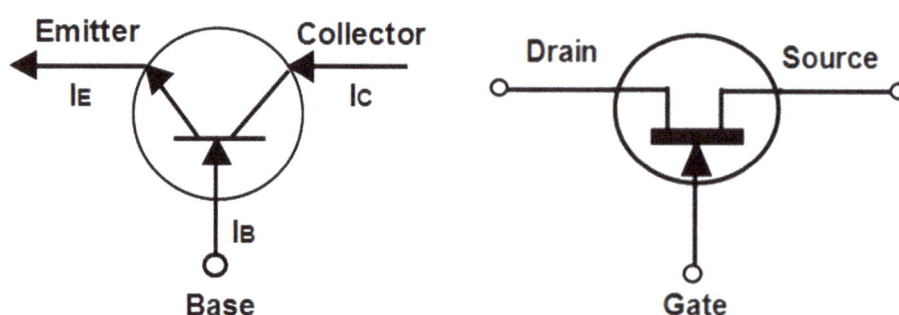

Basically Transistors are electronic switches; instead of operating a toggle a voltage is applied to switch a gate (Base). The diagram above left shows one of the most common Transistors. This is what is designated an NPN Transistor.

The right hand diagram above shows a Junction Field Effect Transistor (JFET) which is a different type of Transistor.

So if the gate is not operating, what is the testing procedure so that the Transistor is not damaged? Where do I connect my Multi-meter and can the voltage of the Multi-meter damage the component? Do I have to remove the Transistor from the Printed Circuit Board to test it or can I do that in place?

So let us have a look at the different types of Transistor, how they work, how to test them and in what circumstances would they be used.

NOTE:

ON ALL ATEX EQUIPMENT FOR HAZARDOUS AREAS SUCH AS ZONES, REMOVAL/CHANGING OF TRANSISTORS FROM A MANUFACTURERS' PRINTED CIRCUIT BOARD CAN ONLY BE CARRIED OUT WITH MANUFACTURER APPROVAL IN WRITING.

FAILURE TO COMPLY WITH THIS MAY BE CLASSED AS AN UNAUTHORISED MODIFICATION AND MAY RENDER ANY CERTIFICATION DOCUMENTATION NULL AND VOID.

ALSO THE DESIGN, MANUFACTURER, TYPE ETC., OF THE TRANSISTOR COMPONENTS MUST NOT BE CHANGED IN ANY WAY.

Transistors:

Bipolar Transistor (BJT) Terminology:

There are two main types of transistor namely: **1)** Bipolar Transistors and **2)** Field Effect Transistors. They both achieve the same objective, but achieve it in different ways. As we go through this section we will discuss both, starting with Bipolar below:

Bipolar Transistor Purpose:

The **'Bipolar Junction' Transistors** are electronic devices which can be used to amplify or switch small electrical currents. As far as switching is concerned, instead of a toggle as in a normal light switch, this transistor uses a small current on the 'base' to do the switching between the emitter and the collector. The 'Bipolar' amplifier side means that a very small current can switch a large one – for instance in a hearing aid.

Shapes:

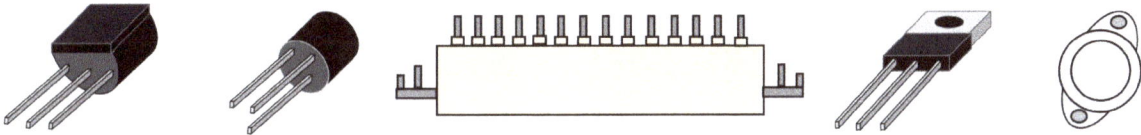

Bipolar transistors come in a range of designs depending upon the make, size and value. I have shown several different shapes above which you may have seen in various electronic circuits. These devices can be divided into three; i.e. a Bipolar transistor has a **Base (Control)**, **Emitter (Output)** and **Collector (Supply)** so the most common devices will have three electrode wires. As in the diagram above they can have several wires.

Different Forms:

So how does this help our explanation? Well **'Bipolar'** transistors are a sandwich of semiconductors, mentioned earlier in the book, and come in two forms i.e. **NPN** (**n**egative, **p**ositive, **n**egative) being the most common, and **PNP** (**p**ositive, **n**egative, **p**ositive). How the different forms work can be found on the next pages now that we have got some of the terminology out of the way.

Doping:

As above transistors are just a sandwich of semiconductors. The semiconductor materials are **'doped'** meaning that an impurity is added to improve the electrical conductivity, i.e. cause the electrons to move more freely, as for instance silicon is an insulator. The **'base'** will be the lightest doped as we want the majority of the current to pass to the **'collector'** not the **'base'**. The doping can be done using materials such as Boron, Aluminium or Gallium.

Bipolar:

The **'Bipolar'** part comes in because we have two types of current; **'P' (Positive)** which is full of **'holes'** and **'N' (Negative)** which is full of **'electrons'**. These can be shown as capitals i.e. **'N'** or lower case **'n'**.

Junctions:

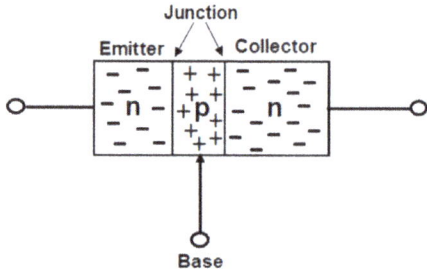

The diagram on the left shows one form of transistor, namely an NPN so as above we have a sandwich of semiconductors with the 'P' (positive) in the middle of two 'N's (negatives). This is called NPN. There are transistors where the 'N' is in the middle of two 'P's and these are called PNP. In the diagram the transistor is at rest or turned off. Where the 'P' meets an 'N' is called a 'PN junction' and as you can see there are two junctions.

NPN & PNP:

The next two pages will go on to explain the differences between **NPN** and **PNP** forms of Transistor, how they work, where they may be used, and an explanation of the most common ones.

The Bipolar NPN Transistor:

The **NPN** Transistor is the most preferred over the **PNP** Transistor; the main reason is the faster mobility of electrons. Below is a very simplified version of how the **NPN** Transistor works; I have not included formulas or very detailed explanations.

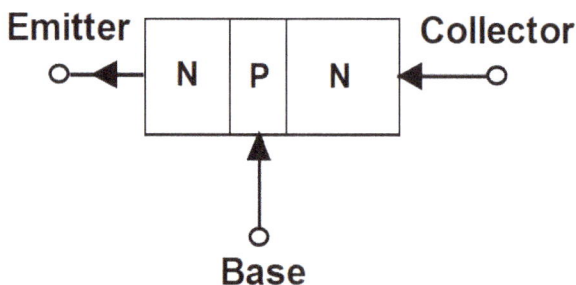

With the **NPN** Transistor we have **two** different silicon or germanium semiconductors ('**N' type**) with easily moved **'Free'** electrons separated by another semiconductor ('**P' type**). The '**P**' type semiconductor has numerous **'holes'** and is much thinner than the '**N**' semiconductors because more current is required at the '**N**' semiconductors and not so much at the base. So this is more of an electron-led unit than the **'PNP'** transistor which is **'hole'** led.

Right is a **NPN** Transistor diagram to match the diagram above. The semiconductor materials are 'doped' meaning that an impurity is added to improve the electrical conductivity, i.e. cause the electrons to move more freely, as for instance silicon is an insulator. The **'base'** will be the lightest doped as we want the majority of the current to pass to the **'collector'** not the **'base'**. The doping can be done using materials such as Boron, Aluminium or Gallium.

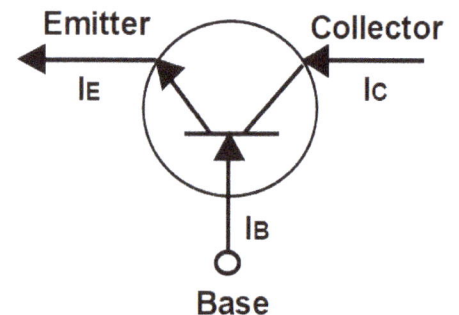

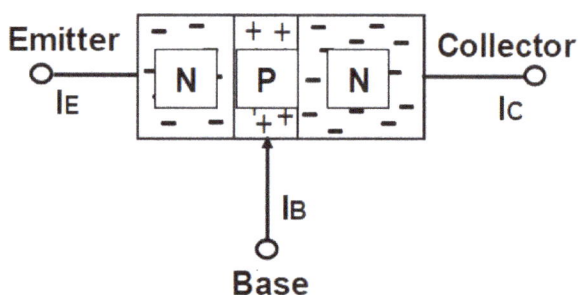

So where are we? Well the diagram on the left shows the **NPN** Transistor at rest or turned off. The '**P**' semiconductor has a positive charge hence the plus signs meaning that there are plenty of **'holes'**, and at either side of the junction are '**N**' semiconductors full of **'electrons'**. So how do we turn this device on and off when we are using it as a switch?

The diagram to the right shows a simple circuit where the Transistor is being used as a switch to operate the lamp. The current flow is from the emitter to the collector via the base. The current limiting resistor is there to limit the base current and in case of any reverse current through the base such as a back EMF. When the contact closes, which could be a microprocessor, the base current flows operating the Transistor and, as above, allowing a larger current to flow from the emitter to the collector via the base, working the lamp. To switch on the Transistor, the base voltage must be higher than the emitter voltage and any volt drop.

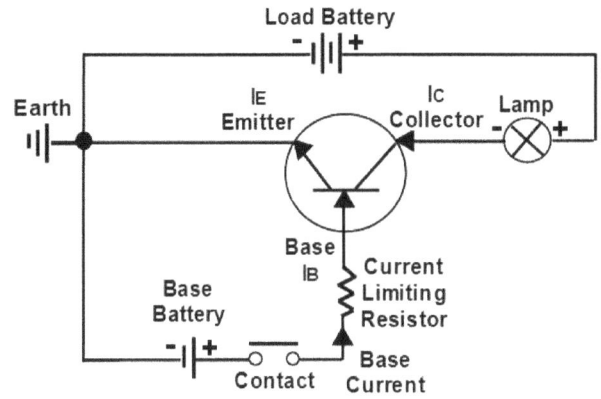

NPN Transistor preferred:

As stated above, the **NPN** transistor is preferred over the **PNP** transistor because of two main reasons:

1) the faster mobility of electrons which in the '**N**' section are highly doped. This may be connected to the term 'carrier mobility'.

2) **NPN** is more suited to the negative/earth system. This may be called a **Common Emitter**. The Transistor acts as an amplifier because small current is put into the input junction (Base) which is forward biased, and the collector, reverse biased, therefore the amplified signal appears at the Collector.

The Bipolar PNP Transistor:

The **NPN** Transistor is the most preferred over the **PNP** Transistor (below); the main reason is the faster mobility of electrons. Below is a very simplified version of how the **PNP** Transistor works; I have not included formulas or very detailed explanations.

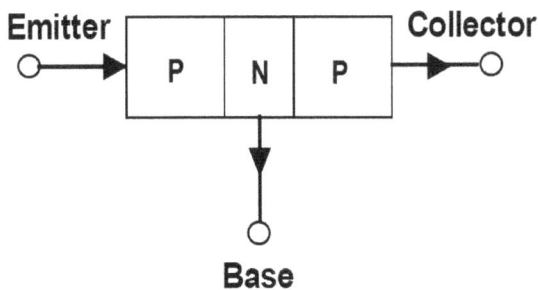

With the **PNP** Transistor we have got **two** different silicon or germanium semiconductors ('**P**' type) with 'holes', separated by another semiconductor ('**N**' type). The '**N**' type semiconductor has numerous 'electrons' and is much thinner than the '**P**' semiconductors.

The semiconductor materials are 'doped' meaning that an impurity is added to improve the electrical conductivity, i.e. cause the electrons to move more freely, as for instance silicon is an insulator. The 'base' will be the lightest doped as we want the majority of the current to pass to the 'collector' not the 'base'. The doping can be done using materials such as Boron, Aluminium or Gallium.

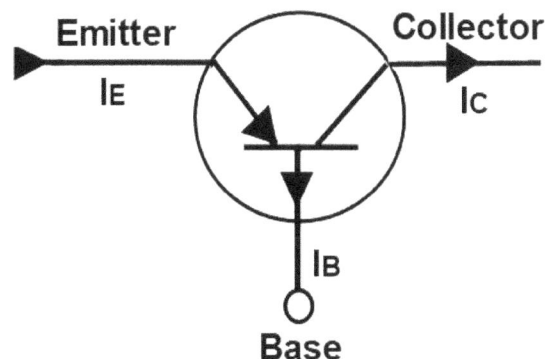

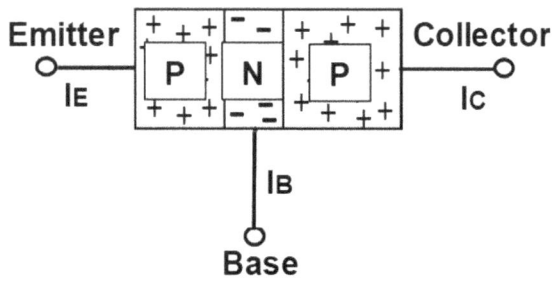

Emitter IE P N P Collector IC

IB

Base

So where are we? Well the diagram on the left shows the **PNP** Transistor at rest or turned off. The **'N'** semiconductor has a negative charge hence the minus signs and is quite thin compared to the **'P'** semiconductors which means that there are fewer **'electrons'** in the centre section and plenty of **'holes'** in the **'P'** semiconductors at either side of the junctions. So how do we turn this device on and off when we are using it as a switch?

The diagram to the right is just a simple circuit where the Transistor is being used as a switch to operate the lamp. The current flow is reversed in this **'PNP'** Transistor compared to the **'NPN'** Transistor. The current limiting resistor is there to limit the base current and in case of any reverse current through the base such as a back EMF. When the contact closes, which could be a microprocessor, the base current flows operating the Transistor and, as above, allowing a larger current to flow from the emitter to the collector via the base thus working the lamp.

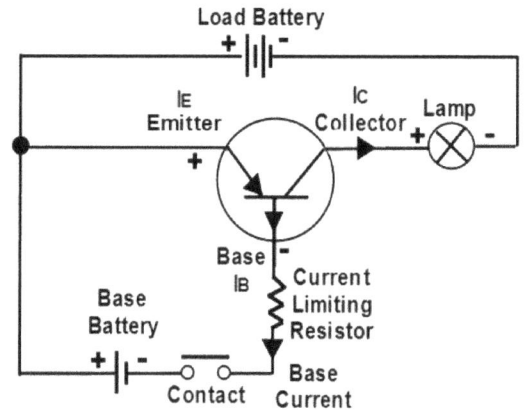

To switch on the Transistor, the emitter voltage must be more negative than the base voltage. As stated above, the **NPN** transistor is preferred over the **PNP** transistor because of two main reasons: **1)** the faster mobility of electrons which in the **'N'** section are highly doped – this may be connected to the term 'carrier mobility' & **2) NPN** is more suited to the negative/earth system. I think the above explanation shows how the Transistor works as a switch and how useful they are in electronic circuits.

The Transistor acts as an amplifier because small current is put into the input junction (Base) which is forward biased, and the collector, reverse biased, therefore the amplified signal appears at the Collector.

Bipolar Insulated Gate Transistor (IGBT) PNP:

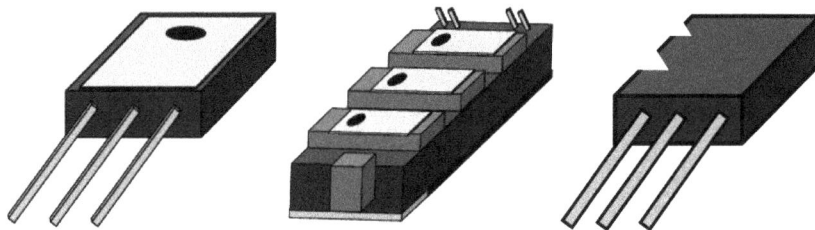

The three electrode Bipolar Insulated Gate Transistors carry more current than a standard Bipolar Transistor. Usually used as an electronic switch they are a cross between a Bipolar Transistor and a Field Effect Transistor. These use a Metal Oxide Material and being also a Field Effect. The Field Effect Transistor gets the abbreviation MOSFET. Look at the Field Effect Transistor MOSFET and you will see how similar they are!

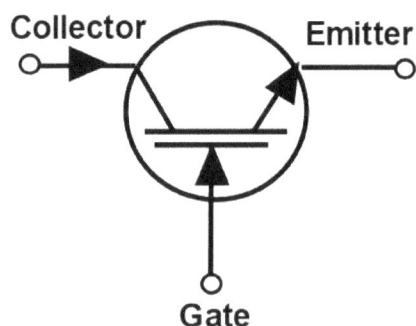

Collector Emitter

Gate

Although this Transistor is a Bipolar unit, which is usually current based, it is controlled by voltage, similar to a Field Effect Transistor so has the best of both worlds with its high impedance input. Because of the insulated Gate there is no input current. Switching speeds with these devices is very fast. When voltage is applied to switch the Transistor on, the current flow is, as in a standard Transistor, from collector to emitter. The fast switching makes this Transistor ideal for Pulse Width Modulation or anything to do with frequency control.

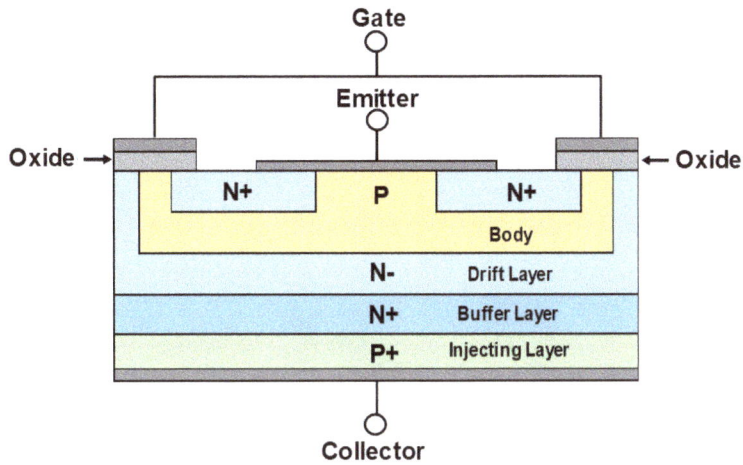

Looking at the diagram above you can see the different **'layers'** within the Transistor and the **'Insulated Gate'**. Remember the current flow is from **'Collector'** to **'Emitter'**; the **'Gate'** just controls that flow and in this case does that with voltage instead of current. Because of the **PN** junction current flow is in one direction so this Transistor cannot be used where a reverse current may be required.

Modes of Operation:

There are three modes of operation: **1)** Forward Blocking mode, **2)** Conduction, & **3)** Reverse Blocking Mode. So remember the Gate, being insulated and at ground potential, looking at the diagram above, for Forward Blocking Mode we put a small forward positive **VOLTAGE** onto the collector terminal. This puts a forward bias between the **N+** and **N-** Drift Layer and a reverse bias between the **P** and the **N-** Drift layer. The transistor is now in **Forward Blocking Mode** awaiting the breakdown voltage.

Now by putting a positive **VOLTAGE** onto the gate we cause electron flow across the formed P & N Junction and as the substrate layers are already in forward bias so we are now in **Conduction Mode**.

Now if we apply a negative voltage to the collector instead of a positive one, the P & N Junction will now be reverse bias and no current flowing. This is called **Reverse Blocking Mode.**

Bipolar Darlington Pair Transistors:

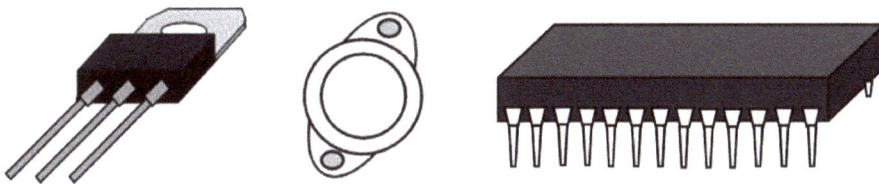

These Transistors, although they are called a 'Darlington Pair', actually come as one Transistor unit containing the two Transistors as shown by a dotted line in the diagram below. Invented by Sidney Darlington in the 1950s, they are 'Bipolar' and can be **NPN** or **PNP** as shown in the diagrams below.

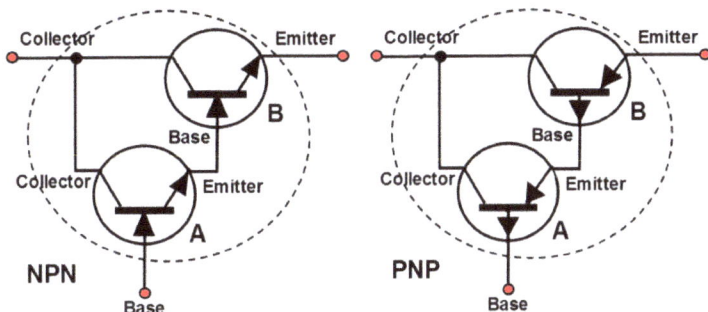

They work on the principle that the current is amplified by the first transistor and that current is compounded to a higher current by the second Transistor. Looking at the diagrams left the 'Emitter' of **Transistor 'A'** is connected to the 'Base' of **Transistor 'B'**. Both Transistor Collectors are joined up and the whole unit would have three terminals i.e. Base, Emitter and Collector coloured red.

One of the major problems with this Transistor is heat due to high current amplification.

Field Effect Transistor (FET) Terminology:

Transistor Purpose:

So what is the difference between Bipolar and Field Effect Transistors? Well both switch and amplify but in different ways.

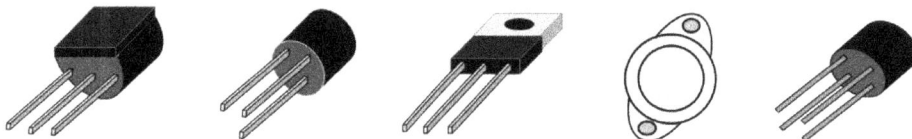

Shapes:

Field Effect Transistors come in a range of designs depending upon the make, size and value. I have shown several different shapes above which you may have seen in various electronic circuits. The most common device will have three electrode wires. As in the diagram above they can have four or more wires.

Different Forms:

These are sometimes called 'Unipolar' Transistors as against 'Bipolar', possibly the most important differences being the 'Bipolar Junction'. Transistors are current controlled and the Field Effect Transistor is voltage controlled. The Bipolar Transistor has a Base, Emitter and Collector; the Field Effect Transistor has a Gate, Drain and Source, just different terminology. Physical shapes are very similar. Field Effect Transistors do not have P & N Junctions.

Field Effect:

It is called a 'Field Effect' Transistor or FET because it actually uses the 'field effect' when operating its control of the output. There are three types of Field Effect Transistor namely 1) Junction Field Effect Transistor 'N' Channel (Electrons); 2) Junction Field Effect Transistor 'P' Channel (Holes); & 3) Metal Oxide Semiconductor Field Effect Transistor (MOSFET) – these types are explained later in this section.

The type of FET using the channels above use an 'N' channel where 'electrons' are used as current carriers OR 'P' channel where 'holes' are used as carriers, but not both at the same time. You will find that the Metal Oxide Semiconductor Field Effect Transistor or MOSFET is the one that is used the most.

The Junction Field Effect Transistor (JFET) 'N' Channel:

Field Effect Transistors have one major difference to Bipolar Transistors in the fact that they use voltage to control a current instead of a small current controlling a larger one. These may be termed **'Unipolar'** Transistors as the control voltage just enables current to flow in the form of electrons in the **'N'** Channel Transistors and holes in the **'P'** Channel Transistors as single carriers.

The input impedance of this Transistor is quite high, whereas a Bipolar Transistor is low. High input impedance causes the input voltage to be more sensitive.

They may look like the diagram above and come in three versions: an **'N'** Channel, a **'P'** Channel and **MOSFET**. Let me try and explain the differences here and how they work compared to a bipolar transistor.

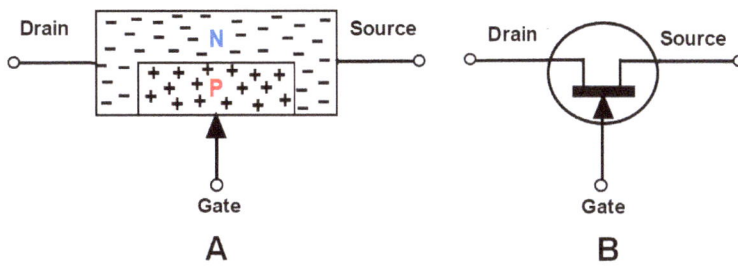

	A		B

As you can see from the diagram **'A'** above there is a complete difference to the bipolar unit. We have a small **'P'** section surrounded by a large **'N'** section. The terminology is also different as we have a Source, Gate and Drain.

The current flow here is from Source to Drain. Remember here we have a positive controlling voltage and that is connected from the Gate to the Source. The current flow is directly through the **'N'** semiconductor from Source to Drain via electrons!

Channels:

The current passing down the **'N'** semiconductor is called a 'Channel'. Looking at the diagram, as mentioned above, there is a large **'N' (Negative)** section which is full of electrons and a small **'P' (Positive)** section that is full of holes.

This may now make a bit more sense of the diagram **'B'** above. So where do we get the name 'Field' from? Well a field is formed when a voltage is applied between the gate and the source, called a 'pinch off' voltage. Varying the voltage between the source and the gate will vary the current flowing through the **'N'** semiconductor.

In this **'N'** channel Transistor the largest section is full of electrons so current flows in the form of electrons. The **'N'** semiconductor is doped with impurities to ensure smooth current flow through the material. The input voltage is proportional to the output current.

ON and OFF:

This is called an **'ON'** Transistor which means that the current path is fully open until a voltage, called a 'pinch' voltage, is applied to either control it or close it, whereas the Bipolar Transistor is called an **'OFF'** Transistor until a current is put on to open or control it.

So in this case, lose the positive pinch off/control voltage and the switch will be fully on. As above, the 'Field' part of the title comes into being because the pinch voltage controls by magnetic field.

The Junction Field Effect Transistor (JFET) 'P' Channel:

Field Effect Transistors have one major difference to Bipolar Transistors in that they use voltage to control a current instead of a small current controlling a larger one. These may be termed **'Unipolar'** Transistors as the control voltage just enables current to flow in the form of electrons in the **'N'** Channel Transistors and holes in the **'P'** Channel Transistors as single carriers.

The input impedance of this Transistor is quite high, whereas a Bipolar Transistor is low. High input impedance causes the input voltage to be more sensitive. Another difference is that this Transistor is an 'on' device until action is taken to control the current, whereas a Bipolar Transistor is an 'off' device until action is taken to switch it on.

They may look like the diagram above and come in three versions an 'N' Channel, a 'P' Channel and a **MOSFET**. Like the Bipolar Transistor, one type is preferred against the other and in this case it is the 'N' channel. Let me try and explain the differences here and how they work compared to a Bipolar Transistor.

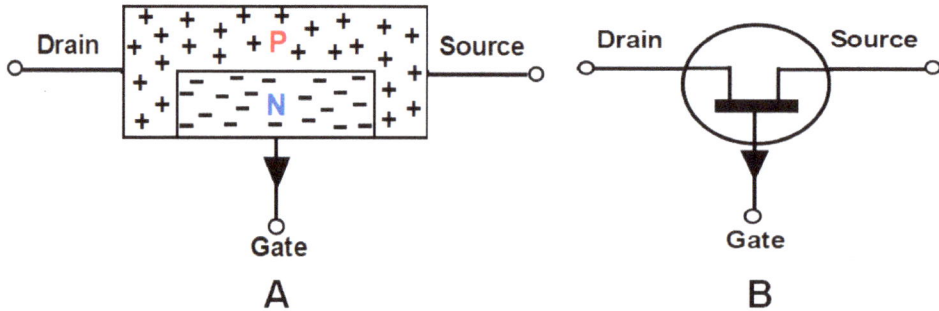

As you can see from the diagram above there is a complete difference to the Bipolar unit. We have a small 'N' section surrounded by a large 'P' section. The terminology is also different as we have a Source, Gate and Drain.

The current flow here is from Source to Drain. Remember here we have a controlling negative voltage and that is connected from the Gate to the Source. The current flow is directly through the 'P' semiconductor.

Channels:

The current passing down the 'P' semiconductor is called a 'Channel'. Looking at the diagram, as mentioned above, there is a large 'P' (**Positive**) section which is full of 'holes' and a small 'N' (**Negative**) section that is full of electrons.

In this 'P' channel transistor the largest section is full of 'holes' so current flows in the form of holes. The 'P' semiconductor is doped with impurities to ensure smooth current flow through the material. Here is where things differ from the 'N' Channel Transistor, with no voltage between the Gate and Source the Channel is wide open so maximum current will flow.

To get control of that current we put a voltage between the Gate and Source called a 'pinch off'' voltage. The input voltage is proportional to the output current.

ON and OFF:

This is called an 'ON' Transistor which means that the current path is fully open until a voltage, called a 'pinch' voltage is applied to either control it or close it, whereas the Bipolar Transistor is called an 'OFF' Transistor until a current is put on to open or control it.

So in this case lose the negative pinch off/control voltage and the switch will be fully on. As above the 'Field' part of the title comes into being because the pinch voltage controls by magnetic field.

Field Effect Transistors (MOSFET):

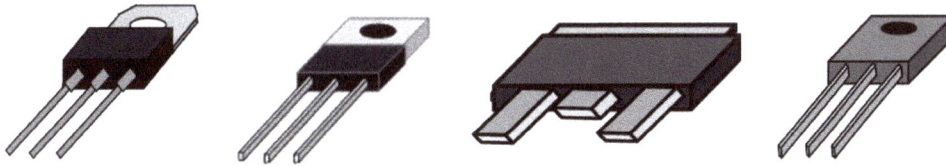

This Transistor has a section that is a different semiconductor in the form of metal oxide, so is called a **M**etal **O**xide **S**emiconductor **F**ield **E**ffect **T**ransistor (MOSFET) it also has an insulated Gate. Examples of what the MOSFET Transistor may look like shown above. The first two have heat sinks.

This is the most complex Transistor to explain. So what are we talking about? Let us look at the diagram below:

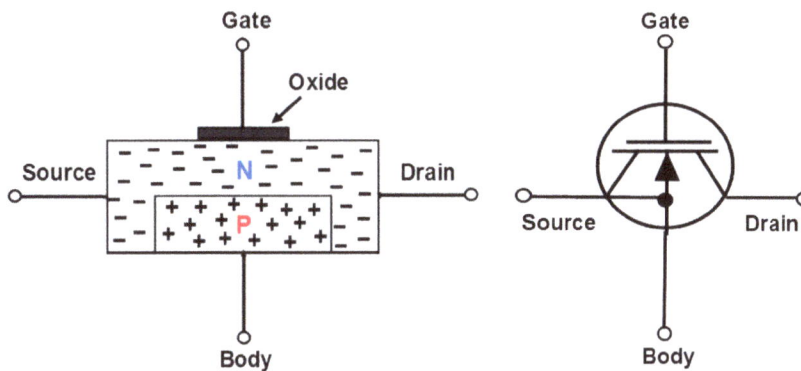

The diagram to the left shows an 'N' (Negative) channel semiconductor which in opposite Transistors can be a 'P' (Positive) channel. These are a three leg device, but some appear to have introduced another leg, besides Source, Drain and Gate, called 'Body' and this is connected to the 'Source'.

Depletion and Enhancement:

This type of Transistor can still be separated further into two different operations. There is **i)** The Depletion type, where a voltage is required at the 'Gate' to turn it **OFF** and **ii)** The Enhancement type where a voltage is required to turn it **ON**. You will note that this type of Transistor is an 'Insulated Gate', the Oxide being the insulator.

So to sum up so far we have a current **'channel'** between the **'Source'** and the **'Drain'** for the current to flow which in an **'N'** channel is 'negative electrons', and a **'P'** channel 'positive holes'. There are two different operations i.e. **'Depletion'** and 'Enhancement'.

In 'Depletion' a voltage is required to turn the current flow **OFF** and in 'Enhancement' a voltage is required to turn the current **ON.** These Transistors do, of course, have a maximum current that they can handle and may destroy themselves should that threshold be exceeded.

Looking at the Transistor diagram to the right, I have drawn the Source connected to the Body as it would be inside of the actual Transistor. I have done this to give you an idea of the end result although standard diagrams may not be drawn exactly like this.

Static Electricity:

These MOSFET Field Effect Transistors are very prone to static electricity – more than other Transistors although remember that the voltage in static electricity can be quite high and hazardous to any electronic components.

In the MOSFET the static can destroy the Gate Oxide and assemblers of circuits containing these Transistors may have to wear antistatic wristbands. In practice an antistatic diode may be inserted to protect the Transistor.

Testing a Transistor:

Testing a Bipolar NPN Transistor: (Using Diode Test Position – Most Efficient Test)

There are three points to test on a Bipolar Transistor: **Emitter, Base** and **Collector**. So how is this done using a Multi-meter? We must develop a method of testing all three points each way as below. Set the Multi-meter to Diode Test.

On very expensive Multi-meters there may be an icon marked **'HFE'** which will have NPN and PNP connections built in. HFE = **H**ybrid parameter **f**orward current gain, common **E**mitter.

The readings of 0.6V below are a nominal example of the approximate test voltage; 0V or OL being the reading of the Multi-meter setting. Obviously there will be no reading Emitter to Collector whether it is an NPN or a PNP Transistor.

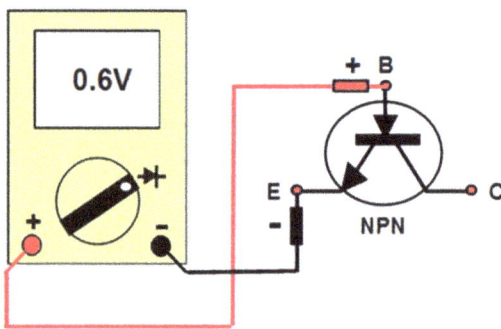

Test 1 - Base (+) to Emitter (-)

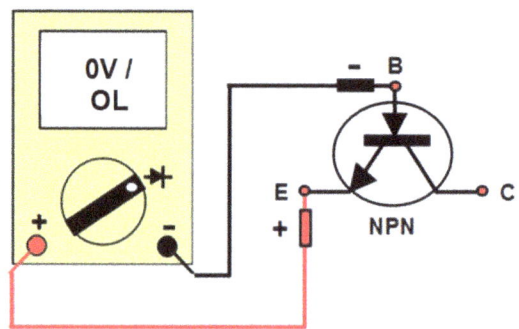

Test 2 - Emitter (+) to Base (-)

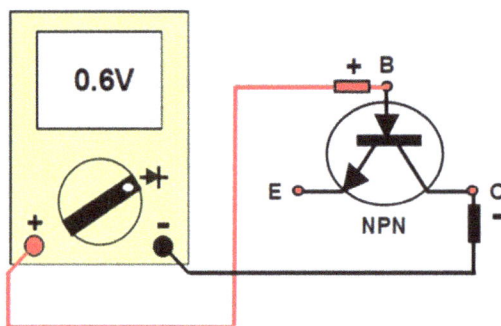

Test 3 - Base (+) to Collector (-)

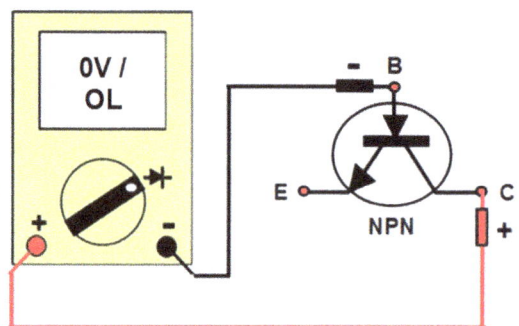

Test 4 - Collector (+) to Base (-)

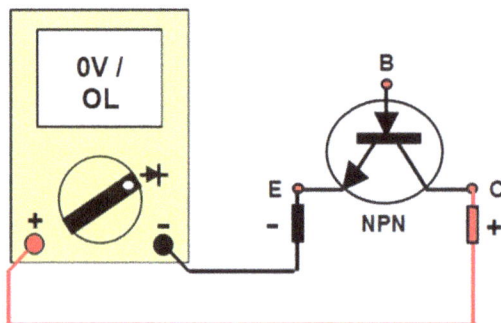

Test 5 - Collector (+) to Emitter (-)

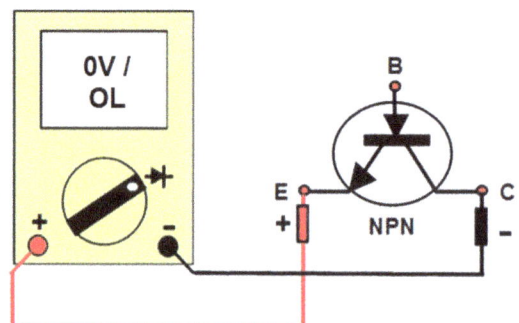

Test 6 - Emitter (+) to Collector (-)

Testing a Bipolar NPN Transistor: Using Continuity Ω Position – Not Recommended as the most efficient Test.

There are three points to test on a Bipolar Transistor: **Emitter**, **Base** and **Collector**. So how is this done using a Multi-meter? We must develop a method of testing all three points each way as below. Set the Multi-meter to Ω **(OHMS)**.

This resistance test is not as accurate as the **Diode Position Test** which is in **VOLTS. If the Multi-meter has not got the Diode Test position then the test would have to be in Ohms.**

The readings of continuity below are a nominal example of the approximate **CONTINUITY RESISTANCE**; OL being 'Over Limit' of the Multi-meter setting. Obviously there will be no reading Emitter to Collector whether it is an NPN or a PNP Transistor.

Continuity is what we are after which would be around the resistance of the leads **0.4Ωish. Tests 2 & 4 should be opposite to the PNP Transistor.**

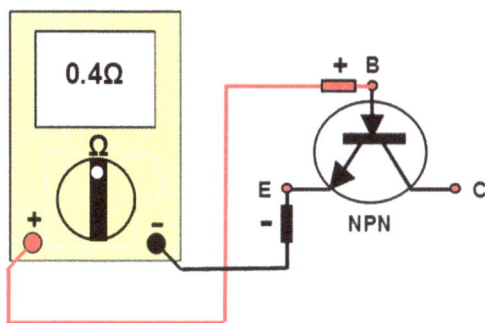

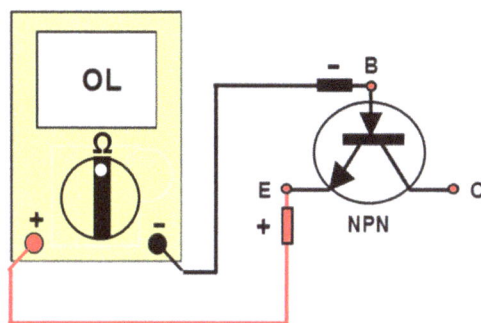

Test 1 - Base (+) to Emitter (-)   Test 2 - Emitter (+) to Base (-)

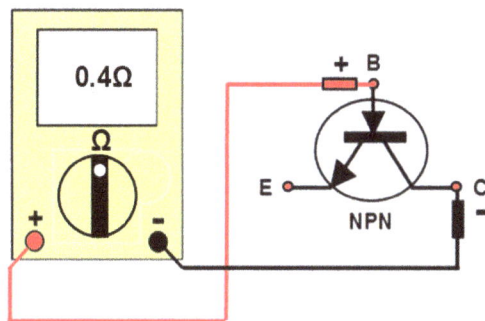

Test 3 - Base (+) to Collector (-)   Test 4 - Collector (+) to Base (-)

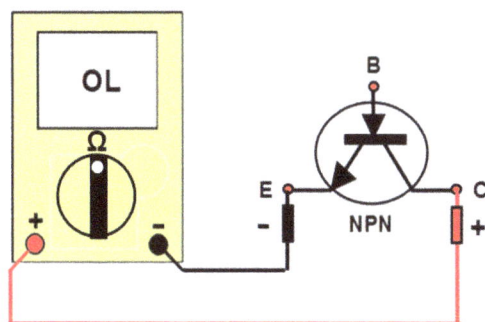

Test 5 - Collector (+) to Emitter (-)   Test 6 - Emitter (+) to Collector (-)

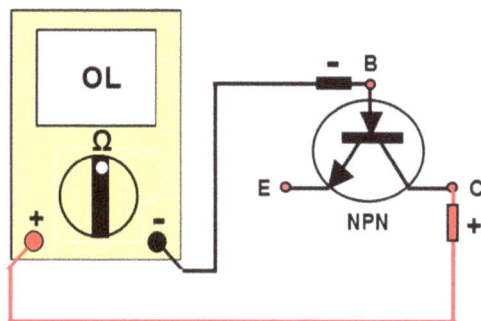

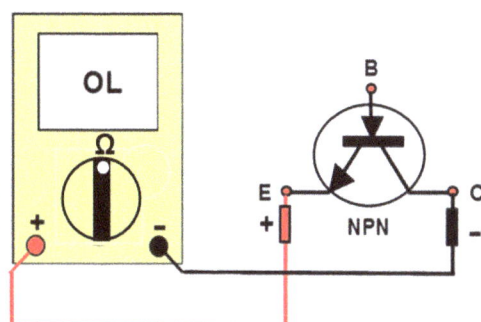

Testing a Bipolar PNP Transistor: (Using Diode Test Position)

There are three points to test on a Bipolar Transistor: **Emitter**, **Base** and **Collector**. So how is this done using a Multi-meter? We must develop a method of testing all three points each way as below. Set the Multi-meter to Diode Test.

On very expensive Multi-meters there may be an icon marked **'HFE'** which will have NPN and PNP connections built in. HFE = **H**ybrid parameter **f**orward current gain, common **E**mitter.

The readings of 0.6V below are a nominal example of the approximate test voltage; 0V or OL being the reading of the Multi-meter setting. Obviously there will be no reading Emitter to Collector whether it is an NPN or a PNP Transistor.

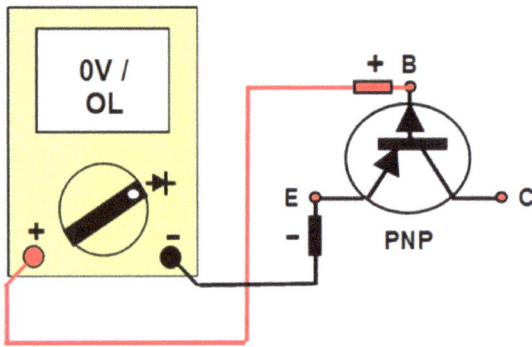

Test 1 - Base (+) to Emitter (-)

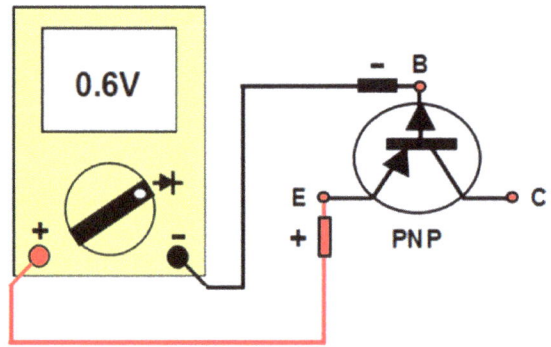

Test 2 - Emitter (+) to Base (-)

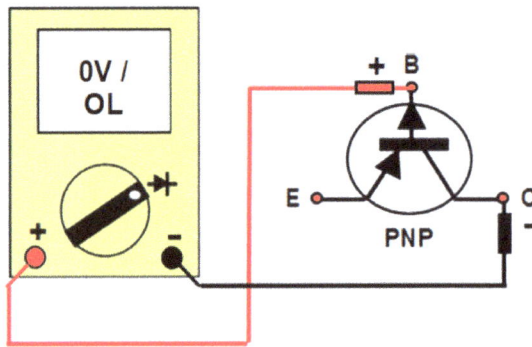

Test 3 - Base (+) to Collector (-)

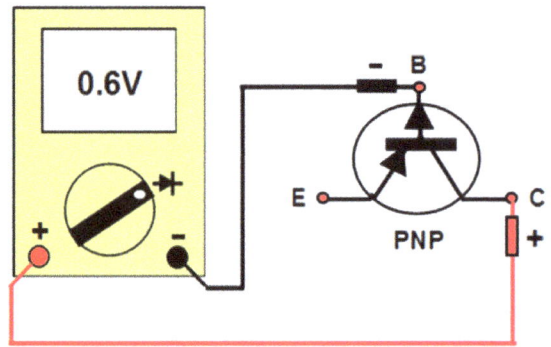

Test 4 - Collector (+) to Base (-)

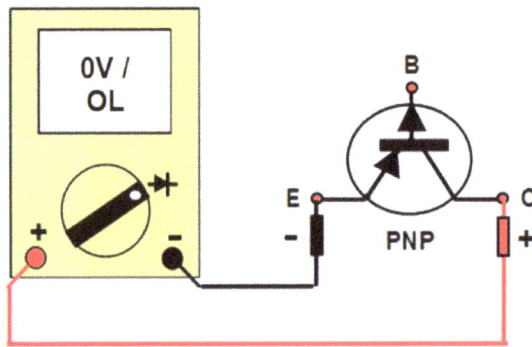

Test 5 - Collector (+) to Emitter (-)

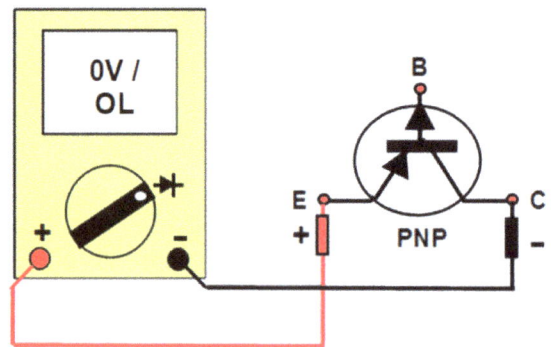

Test 6 - Emitter (+) to Collector (-)

Testing a Bipolar PNP Transistor: Using Continuity Ω Position – Not Recommended as the most efficient Test.

There are three points to test on a Bipolar Transistor: **Emitter**, **Base** and **Collector**. So how is this done using a Multi-meter? We must develop a method of testing all three points each way as below. Set the Multi-meter to **Ω (OHMS)**.

This resistance test is not as accurate as the **Diode Position Test** which is in **VOLTS**. If the Multi-meter has not got the Diode Test position then the test would have to be in Ohms.

The readings of continuity below are a nominal example of the approximate **CONTINUITY RESISTANCE**; OL being 'Over Limit' of the Multi-meter setting. Obviously there will be no reading Emitter to Collector whether it is an NPN or a PNP Transistor.

Continuity is what we are after which would be around the resistance of the leads 0.4Ωish. **Tests 2 & 4 should be opposite to the NPN Transistor.**

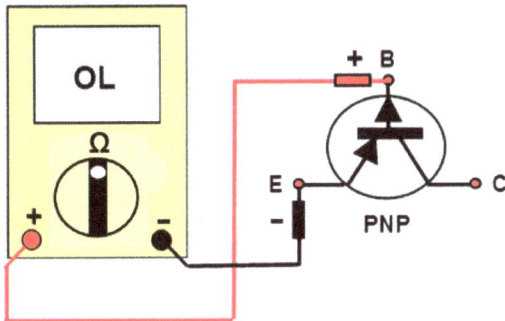

Test 1 - Base (+) to Emitter (-)

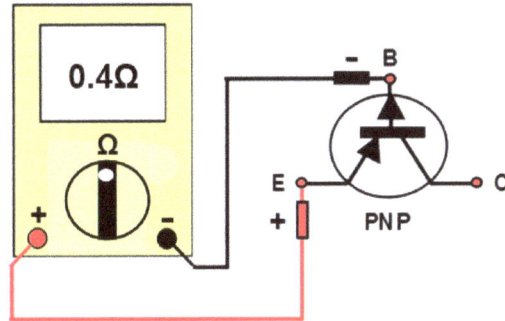

Test 2 - Emitter (+) to Base (-)

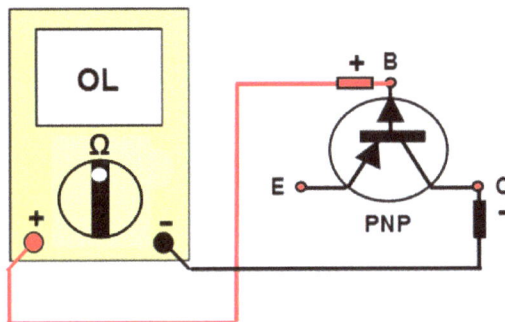

Test 3 - Base (+) to Collector (-)

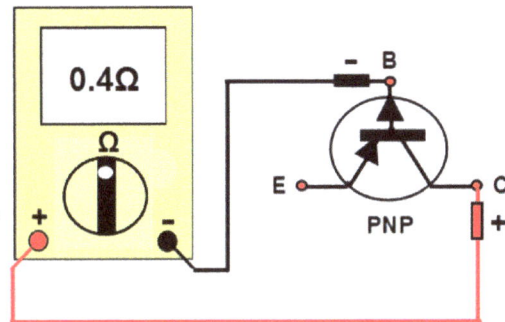

Test 4 - Collector (+) to Base (-)

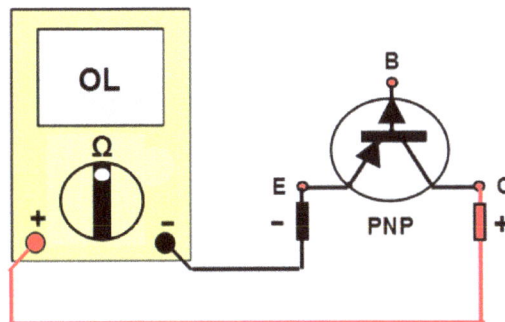

Test 5 - Collector (+) to Emitter (-)

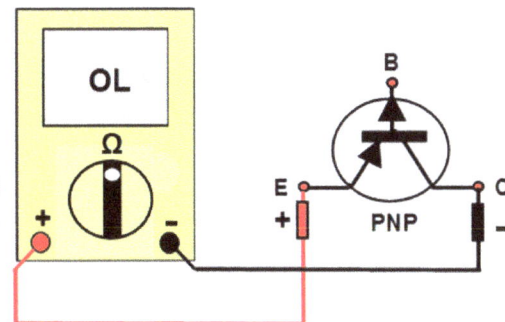

Test 6 - Emitter (+) to Collector (-)

Testing a JFET N-Channel Transistor: (Using Continuity Ω Position)

There are three points to test on a Junction Field Effect Transistor (JFET): **Source**, **Gate** and **Drain**. So how is this done using a Multi-meter? We must develop a method of testing all three points each way as below. Set the Multi-meter in this case to Ohms (Ω).

On very expensive Multi-meters there may be an icon marked **'HFE'** which will have NPN and PNP connections built in. HFE = **H**ybrid parameter **f**orward current gain, common **E**mitter. You can also test this type of Transistor using these terminals if they exist.

Remember on this Transistor we are more or less carrying out a continuity test on the Ohms scale of the Multi-meter so either there is a circuit or there is not. I will just remind you that there may be continuity where there should not be, or no continuity where there should be, hence a faulty Transistor.

Always check the Multi-meter before starting to ensure that the Instrument and its leads are ok. It is also possible for this type of Transistor to store a small charge and thus give you very odd readings that you might not want. Ensure these charges are dissipated before testing!

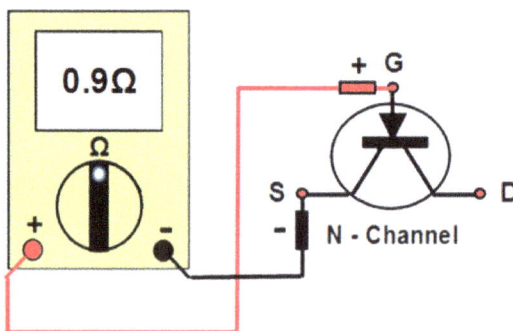

Test 1 - Gate (+) to Source (-)

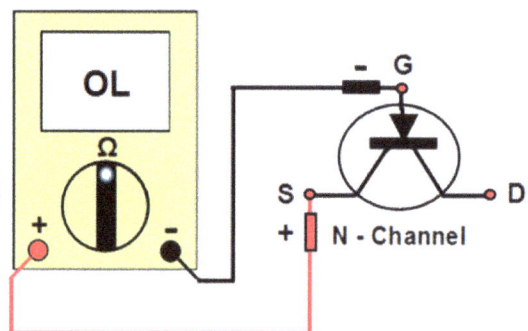

Test 2 - Source (+) to Gate (-)

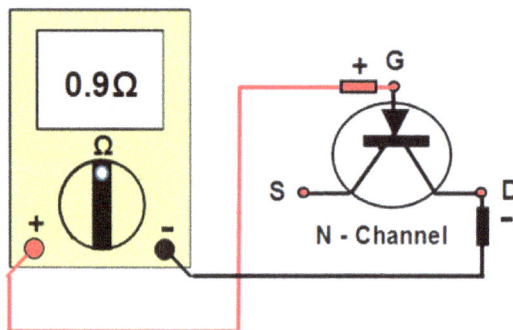

Test 3 - Gate (+) to Drain (-)

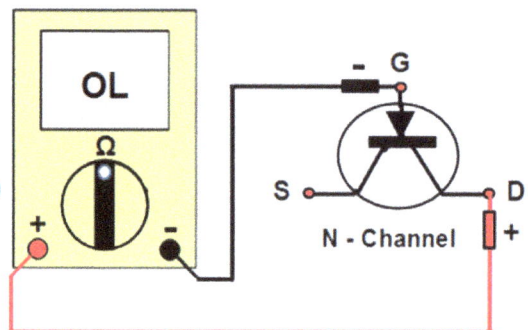

Test 4 - Drain (+) to Gate (-)

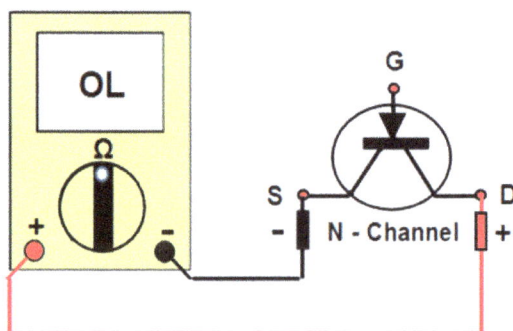

Test 5 - Drain (+) to Source (-)

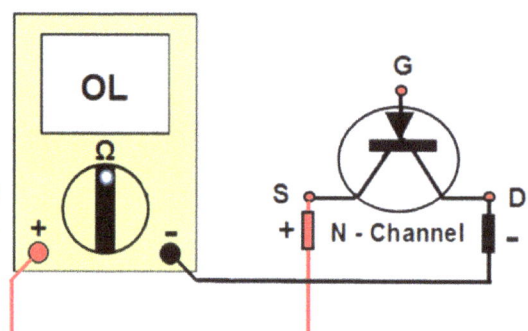

Test 5 - Source (+) to Drain (-)

Polarity of Transistors:

The testing methods on the previous pages will assist you in finding the polarity of various Transistors. On a Bipolar Transistor you are looking for **Base, Emitter and Collector,** and on a Field Effect Transistor or MOSFET you are looking for **Gate, Source and Drain**. On a FET MOSFET you may also be looking for **'Body'.**

On a Bipolar Transistor it must be established if the Transistor is NPN preferred, or PNP. The Transistor will have three electrodes. The first electrode to find is the Base, then, by using previous testing procedures, find the Emitter and Collector.

By far the best and most reliable method would be to enter the Transistor part number into your search engine and see what information it comes up with.

If we take a Flat Sided NPN Bipolar Transistor, the pin configuration on special certain makes or models may have their pins in a different configuration so it is best to check paperwork and test out the polarity yourself to ensure that it is correct. You could destroy the Transistor by getting the polarity wrong.

Some Bipolar metal Transistors called a 'Canister' or 'Top Hat' Transistor have a rim at the bottom with a square tag sticking out, this is usually on the Emitter. The metal case is usually connected to the 'Collector' and in this case, using one of the previous test methods, it should be easy to find the Base.

In a Bipolar Transistor, if you connect the Emitter and Collector the wrong way round the Transistor may still work, but the impedance would be different, possibly higher. This usually is due to higher and lower doping of the Transistor regions. Connecting the Base incorrectly will usually result in the Transistor not working at all.

As above if the Transistor is a Field Effect Transistor then you are looking for a Gate, Source and Drain and in the MOSFET Transistor, also a Body, which is usually connected to the Source. Firstly you need to know if the Transistor is an N-Channel, P-Channel or MOSFET (Metal Oxide Field Effect Transistor).

As stated above, by far the best and most reliable method would be to enter the Transistor part number into your search engine and see what information it comes up with.

Soldering Transistors to a Circuit:

Transistors are very vulnerable to heat and one of the common faults of soldering Transistors to a board or circuit is to cause it to overheat and this is very easy to do without thinking. You will need a heat sink of some kind between where you are soldering to the body of the Transistor and this is easily achieved by adding a small metal crocodile clip just below the body.

Transistor Codes:

Manufacturers put codes on their Transistors to identify one from another i.e. AC100. The code letters below is the European System of Coding for you to check your Transistor:

First Letter:	**Second Letter:**
A – Germanium.	C – Audio Frequency Amplifier.
B – Silicon.	D – Audio Frequency Power Amplifier.
C – Gallium Arsenide.	F – Low Power Radio Frequency Amplifier.
D – Indium Antimonide.	P – High Power Radio Frequency Amplifier.

Any other numbers or letters refer to Manufacturer and model. There are several systems of coding, however, I have just shown one of them above. The first letter indicates the type of Semiconductor used inside of the Transistor, Silicon is probably the most common.

Types of Transistor:

Avalanche Transistor

Ballistic Deflection Transistor (BDT)

Biologically Sensitive Field Effect Transistor (BioFET)

Biomedical and Environmental Transistor

Bipolar Junction Transistor (BJP) NPN

Bipolar Junction Transistor (BJP) PNP

Carbon Nanotube Field Effect Transistor (CNTFET)

Darlington Transistor

Depletion Field Effect Transistor (DEPFET)

Diffusion Transistor

Double Diffusion Metal Oxide Transistor (DMOS)

Dual Gate Metal Oxide Semiconductor Field Effect Transistor (MOSFET)

Electrolyte Oxide Semiconductor Field Effect Transistor (EOSFET)

Fast Recovery Field Effect Transistor

Fast Reverse Field Effect Transistor

Fin Field Effect Transistor (FinFET)

Fast Recovery Epitaxial Field Effect Transistor (FREDFET)

Graphene Nanoribbon Field Effect Transistor (GNRFET)

Heterojunction Bipolar Transistor (HBT)

Heterostructure Insulated Gate Field Effect Transistor (HIGFET)

Hexagonal Field Effect Transistor (HEXFET)

High Electron Mobility Transistor (HEMT)

High Frequency Transistor

Injection Enhanced Gate Transistor (IEGT)

Insulated Gate Bipolar Transistor (IGBT)

Insulated Gate Field Effect Transistor (IGFET) Dual Gate 'N' Channel

Insulated Gate Field Effect Transistor (IGFET) Dual Gate 'P' Channel

Insulated Gate Field Effect Transistor (IGFET) 'N' Channel

Insulated Gate Field Effect Transistor (IGFET) 'P' Channel

Insulated Gate Field Effect Transistor (IGFET) Non Punch Through

Insulated Gate Field Effect Transistor (IGFET) Punch Through

Inverted-T Field Effect Transistor (ITFET)

Ion-Sensitive Field Effect Transistor (ISFET)

Junction Field Effect Transistor (JFET) 'N' Channel

Junction Field Effect Transistor (JFET) 'P' Channel

Metal Nitride Oxide Semiconductor Transistor (MNOS)

Metal Oxide Semiconductor Field Effect Transistor (MOSFET)

Metal Semiconductor Field Effect Transistor (MESFET)

Multiple Base Transistor

Multiple Emitter Transistor

Nanoparticle Organic Memory Field Effect Transistor (NOMFET)

N Channel Metal Oxide Semiconductor Field Effect Transistor (MOSFET)

P Channel Metal Oxide Semiconductor Field Effect Transistor (MOSFET)

Negative Metal Oxide Semiconductor Transistor (nMOS)

NPN Transistor

Organic Field Effect Transistor (OFET)

Photo Transistor

PNP Transistor

Point Contact Transistors

Power Transistor

Quantum Field Effect Transistor (QFET)

Schottky Transistor

Small Signal Transistor

Small Switching Transistor

Sziklai Pair Transistor

Thin Film Transistor (TFT)

Trench Metal Oxide Semiconductor Transistor (TMOS)

Tunnel Field Effect Transistor (TFET)

U-groove Metal Oxide Semiconductor Field Effect Transistor (UMOS)

Unijunction Transistor (UJT) 'N' Base

Unijunction Transistor (UJT) 'P' Base

Vertical Metal Oxide Semiconductor Transistor (VMOS)

Vertical Slit Field Effect Transistor (VeSFET)

Bipolar Transistor: Questions and Answers

Q1 – Is a Bipolar Transistor a switch?

A1 – **YES** a Bipolar Transistor basically is an electronic switch, but instead of a toggle a small current is used at its base to switch it on and off.

Q2 – Is a Bipolar Transistor an amplifier?

A2 – **YES** the Bipolar Transistor acts as an amplifier because small current is put into the input junction (Base) which is forward biased and the collector reverse biased therefore the amplified signal appears at the Collector.

Q3 – Which are preferred, **NPN** or **PNP** Transistors?

A3 – **The NPN TRANSISTOR** is the most common due to faster mobility of electrons.

Q4 – Can **NPN** Transistors be changed for **PNP** in an electronic circuit?

A4 – **NOT EASILY** there are polarity differences which need to be overcome.

Q5 – What is 'doping'?

A5 – **ADDED MATERIALS** used in the 'N' and 'P' which might be Silicon or Germanium are insulators, and the doping is adding a material such as Boron, Aluminium or Gallium to make them more conductive. The base is doped lightly, the collector is doped moderately and the emitter is doped heavily.

Q6 – Why are some Transistors called Bipolar?

A6 – Because it uses two different polarities namely 'N' semiconductor carries electrons and 'P' semiconductor contains holes.

Q7 – What are junctions?

A7 – The point where one semiconductor meets a different one i.e. 'P' meets an 'N'.

Q8 – Does a Bipolar Transistor operate on voltage or current?

A8 – **CURRENT** so this Transistor is called a 'Current Controlled Device'.

Q9 – Which direction does the current pass?

A9 – **ELECTRONS** try to pass from emitter to the collector (negative to positive) so current flows from the collector to the emitter (positive to negative). The arrowhead shows current direction.

Q10 – Which type of Transistor was invented first?

A10 – **BIPOLAR** only by a few years around the late 1940s.

Q11 – What are Semiconductors?

A11 – **NOT INSULATORS** but neither are they straight conductors. They allow a certain amount of electrons to flow through them. Explained earlier in the book.

Q12 – Are Transistors marked as to their polarity?

A12 – **SOME ARE.** You may find that the legs are numbered. The way to find this information is to look at the datasheets supplied by the manufacturer.

Field Effect Transistor Questions and Answers:

Q1 – Is a Field Effect Transistor a switch?

A1 – YES a Field Effect Transistor basically is an electronic switch, but instead of a toggle a small current is used at its base to switch it on and off.

Q2 – Is a Field Effect Transistor an amplifier?

A2 – YES the Field Effect Transistor acts as an amplifier because small current is put into the input junction (Base) which is forward biased, and the Collector, reverse biased, therefore the amplified signal appears at the Collector.

Q3 – Are there **NPN** or **PNP** Field Effect Transistors?

A3 – FIELD EFFECT TRANSISTORS DO NOT HAVE PN JUNCTIONS!

Q4 – Why are some Transistors called Unipolar and others Bipolar?

A4 – PN Junction Transistors are Bipolar and Field Effect Transistors are Unipolar.

Q5 – Does a Field Effect Transistor operate on voltage or current?

A5 – VOLTAGE so this Transistor is called a 'Voltage Controlled Device'

Q6 – Which direction does the current pass?

A6 – ELECTRONS try to pass from Drain to the Source so current flows from the Source to the Drain. The arrowhead shows current direction.

Q7 – Which type of Transistor was invented first?

A7 – BIPOLAR only by a few years around the late 1940s.

Q8 – What are Semiconductors?

A8 – NOT INSULATORS but neither are they straight conductors. They allow a certain amount of electrons to flow through them. Explained earlier in the book.

Q9 – Are these Transistors marked as to its polarity?

A9 – NOT USUALLY on the actual Transistor.

Q10 – Why are these transistors called Field Effect?

A10 – Because the DRAIN output is controlled by a 'Field' at the depletion zone of the Gate.

Q11 – Is there more than one type of Field Effect Transistor?

A11 – YES there is an 'N' Channel and a 'P' Channel.

The Diode & Rectifier Section:

Diodes are one of the most common components on Printed Circuit Boards and we take them for granted. We do not bother about what task these components have in the scheme of the Printed Circuit.

We have discussed in an earlier section "what is a PN Junction?" Well now in the case of our Diode we are going to see where it fits in with this component. Below I have fitted the PN Junction diagram into the diagram of a Diode just to give you an idea of how the Diode and in fact other components work.

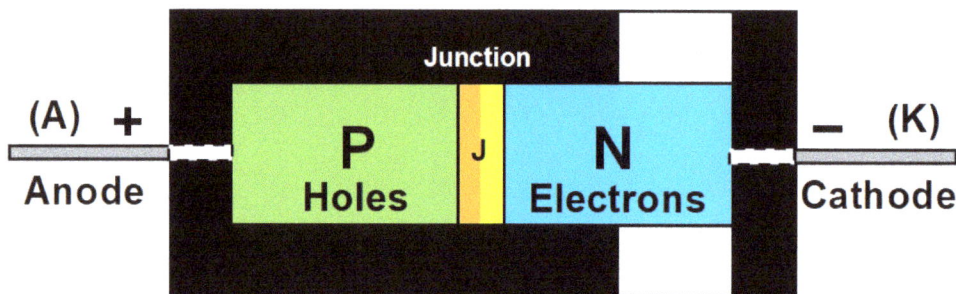

Like the Transistor and Resistor there are so many different types of Diode. I have explained some of the more common Diodes in the next few pages and listed many of the different types.

On some types of Diode you will find that there is a volt drop of 0.7V due to forward resistance and breakdown voltage. Ideally it **should** be 0V.

We will discuss what the difference is between a Diode and a Rectifier. In modern times that would be not a great deal but in the past Rectifiers were large units with plates.

How can we test a Diode on a Printed Circuit Board? Do we have to remove the Diode in order to test it? Are there any special sections of the Multi-meter for specifically testing Diodes?

We look at the P & N Junction, an example above, and how current is transmitted through the components:

NOTE:

ON ALL ATEX EQUIPMENT FOR HAZARDOUS AREAS SUCH AS ZONES, REMOVAL/CHANGING OF DIODES FROM A MANUFACTURERS' PRINTED CIRCUIT BOARD CAN ONLY BE CARRIED OUT WITH MANUFACTURER APPROVAL IN WRITING.

FAILURE TO COMPLY WITH THIS MAY BE CLASSED AS AN UNAUTHORISED MODIFICATION AND MAY RENDER ANY CERTIFICATION DOCUMENTATION NULL AND VOID.

ALSO THE DESIGN, MANUFACTURER, TYPE ETC., OF THE DIODE COMPONENTS MUST NOT BE CHANGED IN ANY WAY.

Diodes & Rectifiers:

Diodes & Rectifier Introduction: These are 'passive' components which means that they consume energy and do not produce it. They are very important items in electronic circuits and probably, alongside resistors, one of the most common that you will come across.

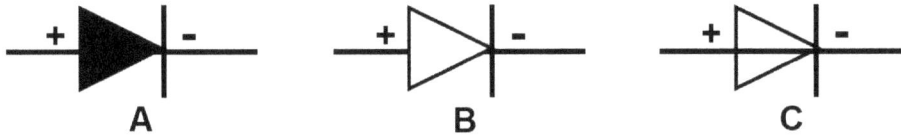

Diode Outline: Let us first look at the Diode, drawing symbols above, these are most frequently used to change AC to DC but some can be used to block undesirable currents such as back EMFs or be used as a static switch.

When we are designing our circuit we need to know exactly which type of Diode we need to complete our task, and there are many as you will see in this section.

Although the drawing symbols can be any of the above, I have drawn Diodes in this section as top left '**A**' (I.E.E.E.).

Usually we can say that a Diode is a semiconductor electronic device with two electrodes, an 'Anode' (+) and a 'Cathode' (-), and it converts AC to DC, but there are others which carry out a totally different function.

The difference between the types of Diode is what happens inside its body. So how, using a Diode, do we provide very vulnerable, important circuits, especially Intrinsic Safety Loops, with protection from voltages that are higher than safety would desire? We will discuss this as we go.

If I wired this Diode the wrong way round would it work? With, say, a Standard Diode, of course not; the current would never flow against the arrow because going back to our P & N Diagram, when wired correctly, forward bias, there would be no, or a very small, depletion zone so much less resistance and current would flow quite easily.

Wired incorrectly, reverse bias, the depletion zone would increase, making the 'breakdown' required impossible for current to flow.

A Zener Diode will allow current to flow against the arrow head if it reaches a certain value, so it is meant to be wired in reverse bias so it can carry out its protection. We will discuss later how it does that. To obtain the correct Diode firstly check the Diode number, usually on the side, and the number of pins.

Rectifier Outline: Also in this section we talk about the many different types of Rectifier. These also change AC to DC and achieve many of the same objectives as Diodes.

When we talk about Rectifiers to an older electrical person, such as myself, an imaginary, fairly large item with many square plates connected together comes to mind and although these still exist, I can assure you that is mostly not the case in modern day.

A Rectifier can be a standalone electronic unit; in many cases we can say that a Rectifier is in actual fact a whole electronic circuit that **MAY**, even, contain a Diode or Thyristor as part of it.

There are two main types of Rectifier:

1) **'Controlled'** which is, say, where we have a Thyristor or a SCR: **Silicon Controlled Rectifier**. Thyristors are explained in a later section of their own.

2) **'Uncontrolled'** where we just have a Diode.

We will also look at how Thyristors & SCRs fit into our electronic circuits and how closely they are related to Diodes and Rectifiers. A Stand-alone Rectifier will be marked anode (+) and/or cathode (-) somewhere.

The Standard Diode Description: Let us firstly look at a **Standard Diode** and see how it works before we go much further. This should give us an idea of how semiconductor devices work. Earlier we talked about the purpose of a P & N junction and what happens, well now we will see where these junctions fit in with actual units.

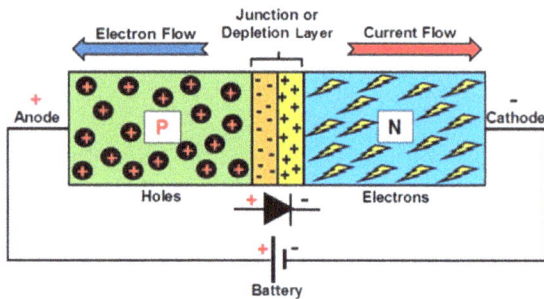

Previously we have looked at the P & N Junction and how forward and reverse bias can affect the working of the diode to our advantage or disadvantage. Looking at the P & N diagram, left, we can see that the green section 'P' is full of positive holes and the blue section 'N' is full of negative electrons. Several positives & negatives will move immediately to form the yellow junction. This P & N is connected in forward bias, battery positive to 'P' and negative to 'N'.

Usually a Standard Diode would have the appearance of the diagram above and if you look on electronic circuit boards you may find that this is the most common type of Diode. The Standard Diode is a two electrode device and has a black body marked at one end with a silver ring called the 'Band' and this is the Cathode (-), marked '**K**' on drawings as '**C**' is used for '**C**ollector'. On the symbol the vertical line at the tip of the arrowhead would be the silver ring on the Diode! On a 'glass' Diode this may be a black spot.

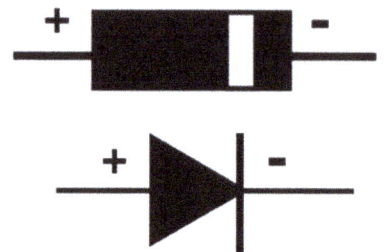

The current path here is forward bias as per the direction of the arrowhead, positive to negative and cannot return against the arrowhead negative to positive.

With a Standard Diode there is no intended breakdown voltage, i.e. the voltage required to 'push' the current against the arrowhead, if the Diode was connected in reverse bias. A component will break down if the voltage becomes high enough anyway, but would of course destroy this Diode.

This Diode 'switches on' from being dormant at a certain voltage value and this is when the voltage at the Anode (+) is higher than the voltage at the Cathode (-). When Diodes pass current it is all to do with electrons finding 'holes' across the P & N Junction above which determines when current can flow and when it cannot. Electron flow is of course opposite to current, negative (-) to positive (+).

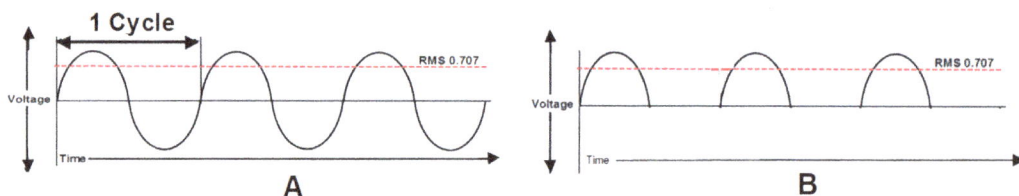

Let us look at a simple analogy. The diagram '**A**' above shows a sine wave created by an AC voltage; you will see that 0.707 is marked as a red line. 0.707 is the **Root Mean Square Value** that all instruments read and is the most efficient part of the wave, the DC equivalent of the AC Sine Wave if you like, which is **NOT THE PEAK**.

This would be ideal DC, but with a Diode or Rectifier on its own, we will not reach the straight line DC that we would like. If we put a standard diode into the circuit it would only allow the positive part of the sine wave through in the direction of the arrow and would not allow the negative to come back so we would end up with a '**Half Cycle**' wave form like the diagram '**B**' above. This is called '**Half Wave**' Rectification.

We are nearer to the straight line, which is where we want to be, but our DC is now what is called very noisy in terms of voltage and useless large gaps between peaks. In an analogy with liquid in pipes, many diodes such as the Standard Diode are what you might class as a **ONE-WAY VALVE**. As we go on you will see that we can use certain Diodes and Rectifiers to produce what is called '**Full 'Wave** Rectification which will be even closer to our straight line.

I have selected several common Diodes to describe as we go through this section; many of these units were in Instruments and electronic circuits where I worked. Several Diodes may overlap with each other and where they are used. We will also look at the working of a Thyristor and see how closely they are connected to the Diode family. Later I have listed many not so common Diodes.

Avalanche Diode Description:

A Protection Diode. Mainly used like the Zener Diode as a Protection Diode, protecting delicate equipment from unwanted or surges in voltage!

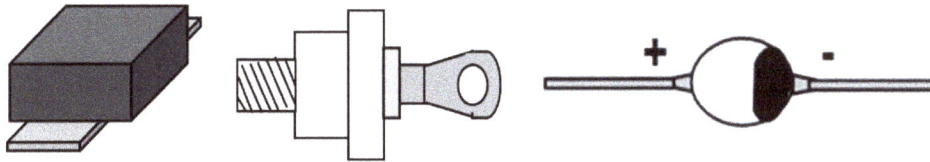

This Diode has many design shapes, three of which I have shown above. They can be made of glass and would be two electrode units. Avalanche Diodes might not be as common as the Zener Diode, which we will talk about later, but like any other Diode has an anode(+) and a cathode(-).

You will notice that the symbol to the right for the Avalanche Diode is exactly the same as that for a Zener Diode, both of which are sometimes called a Breakdown Diode.

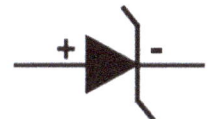

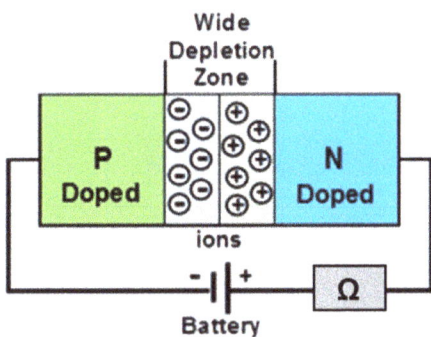

This is usually a Silicon semiconductor Breakdown Diode that is connected in **reverse bias** as per the diagram to the left. This Diode can break voltages higher than a Zener Diode because of a wider P & N depletion zone and not being so highly doped.

Look how the battery is connected in relation to the P & N sections on the left. Providing the voltage stays within the 'Breakdown Parameters' then current will not flow, but if the voltage exceeds the 'Breakdown' value then current will flow against the arrow head, unless very excessive, without damage to the Diode.

You will notice in the above P & N structure that the depletion zone is made up of what are called positive immobile ions on the 'N' side and negative immobile ions on the 'P' side. The centre of which where the immobile ions meet is called the 'junction'. The P & N regions are doped, meaning another substance is added to the semiconductor to assist electron flow.

Diagram to the right shows Avalanche Diode connected in reverse bias in parallel with the equipment to be protected. The red arrow shows the direction of fault or transient voltage at breakdown. The voltage would go through the Diode to earth and blow the fuse. It is that simple.

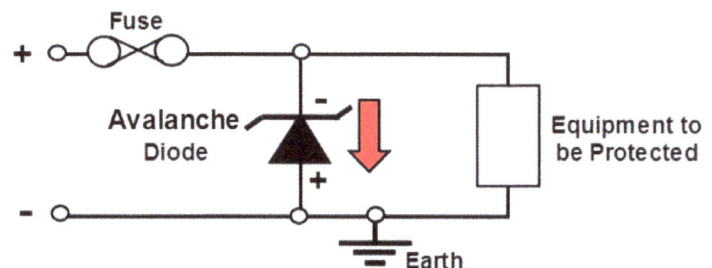

The Diode will have a breakdown voltage value. At the 'Breakdown' value, as the atomic particles move very fast they collide with other atoms, breaking their covalent bonds, knocking more and more particles from their orbits so the energy increases exponentially, and thus we get the **Avalanche Effect**!

There are several uses for an Avalanche Diode, but it is mostly, like the Zener Diode, used to protect delicate electronics from rises in voltage. Although this Diode is meant to be connected in reverse bias as per the P & N Diagram, it will of course work in forward bias and operate like any other Diode.

Avalanche Photodiode Description:

Used for Optical communication and accurate distance measurement. Suitable for high frequency.

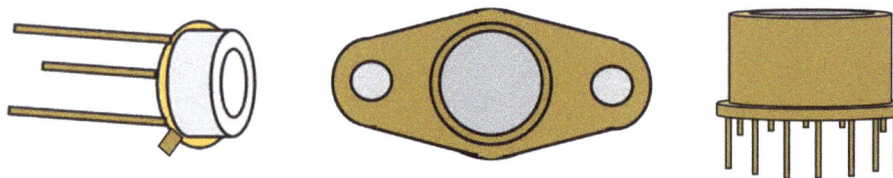

Avalanche Photodiode shapes are very limited; usually they are shaped as the above diagram and have from two to multi-electrodes. Very similar in looks to the Standard Photodiode.

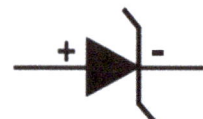

The symbol for the Avalanche Photodiode is the one to the right which is the same as an Avalanche Diode and a Zener Diode. Used, for instance, in fibre optics where light has to be changed to electrical energy.

Avalanche Photodiode semiconductor material can be Silicon (Si), Germanium (Ge) or Indium Gallium Arsenide (InGaAS) in standard Diodes. Gallium Nitride (GaN) semiconductor when used in ultraviolet light and Mercury Cadmium Telluride (HgCdTe) for infrared.

They are highly sensitive semiconductor Diodes, very similar to the Optodiode, PIN Diode or PN Diode. One of the main disadvantages is that they require a higher operating voltage.

The objective of this Diode is to convert light photons into electrical energy. This Avalanche Photodiode is a bit more sensitive than a PIN Diode which is described later in this section.

As you can see by the diagram below they operate in reverse bias, cathode (+) to the 'N' region. Their response time is very high compared to the similar PIN Diode.

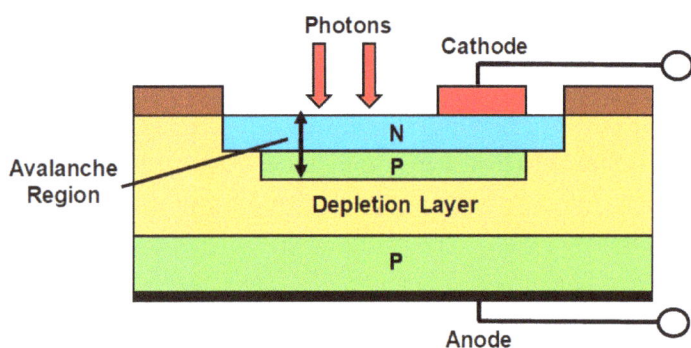

So looking at the P & N diagram to the left there are two doped 'P' regions and one doped 'N' region; photons fall onto the 'N' region. This diode is connected, with a higher voltage, in reverse bias and works similarly to the earlier Avalanche Diode. So just to recap, an avalanche is very fast electrons bumping into other atoms, forcing more and more electrons from their orbit, and those go into other atoms and so the avalanche grows.

Axial Lead Diode Description: Any diode where the leads form an 'axis' through the Diode.

The axial leads on the Diodes above are usually fairly long so as to allow easier fixing and positioning on a printed circuit board. These components can also be obtained as high voltage (2-20kV) Diodes.

Sometimes with a glass body such as the above left hand diagram, HV Axial Diodes will be much more robust and electrode wires thicker than their standard counterparts.

Diodes, Rectifiers and Thyristors:

Diode Bridge Description: Can be known as a Bridge Rectifier or Graetz Bridge and there are several drawing symbols but I have used only one. This of course could be an example of an Uncontrolled Rectifier as only Diodes are used.

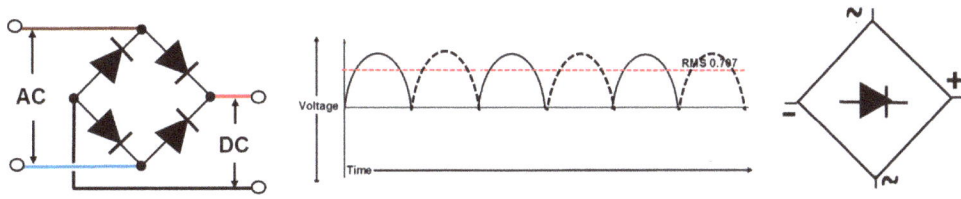

By using what is known as a Diode Bridge, above left, we can achieve what is known as Full Wave Rectification. We have actually inverted the negative part of the sine wave, so now our DC is getting nearer to a straight line and by using capacitors we can make it even better.

Light Emitting Diode (LED) Description: The first LEDs were red only but these days the natural colour of the LED depends upon the semiconductor material and the symbol for an LED is shown on the right. There are several types of **LED:** Alphanumeric, lighting, miniature, high powered etc. The symbol for an **LED** in a circuit is shown below.

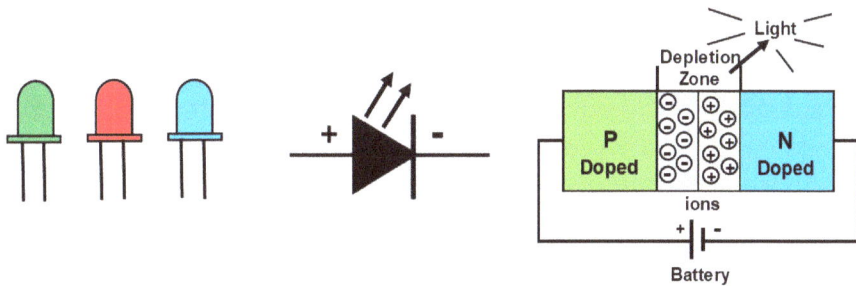

These Diodes convert electrical energy into light. They are found in numerous applications these days such as TV remote controls etc. Several complex semiconductor materials used in the making of LEDs are Aluminium Gallium Arsenide, Aluminium Indium Gallium Phosphide and Indium Gallium Nitride.

This is a rather unusual P & N Diagram as no other Diode produces light so I thought I would show what happens at P & N level. Remember the P & N Semiconductor material doping depends upon the way the electrons flow.

We take a heavily doped semiconductor material and, when in forward bias as the diagram above, free electrons pass into holes at the PN Junction they emit light, instead of heat, as photons. We call this unique process of electron flow **'electroluminescence'.**

Selenium Rectifier Description: This piece of equipment is a Rectifier not a Diode.

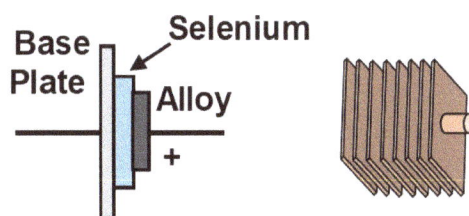

They have been around since the late 19th century. Older electrical guys like myself will remember these well as square plated and in high current circuits such as large battery chargers etc. The semiconductor Selenium is sandwiched between an alloy positive plate and the base plate. Before Selenium came Copper Oxide, but the **Silicon Diode Rectifier** has now replaced them.

Zener Diode Description:

May be called Voltage Regulator Diodes or Breakdown Diodes. There are many diodes that can do other completely different tasks and they can be used to protect delicate electronic components, especially Intrinsically Safe Circuitry in Hazardous Areas.

We have here a heavily doped semiconductor device that is designed to protect a circuit being connected in reverse bias.

Anyone who works with instruments in hazardous areas will have come across the **Zener Diode** and it may be a **Glass Diode** shaped and coloured as the diagram above, or inside an interface unit called a **Zener Barrier**

The drawing symbols for these Diodes can be either of the examples above. The black ring will indicate the cathode. Looking at the diagram above which is a simple **Zener Barrier** protecting an **Intrinsically Safe** piece of equipment in a hazardous area, you will see that the **Zener Diodes** which would be inside the barrier are connected in reverse bias.

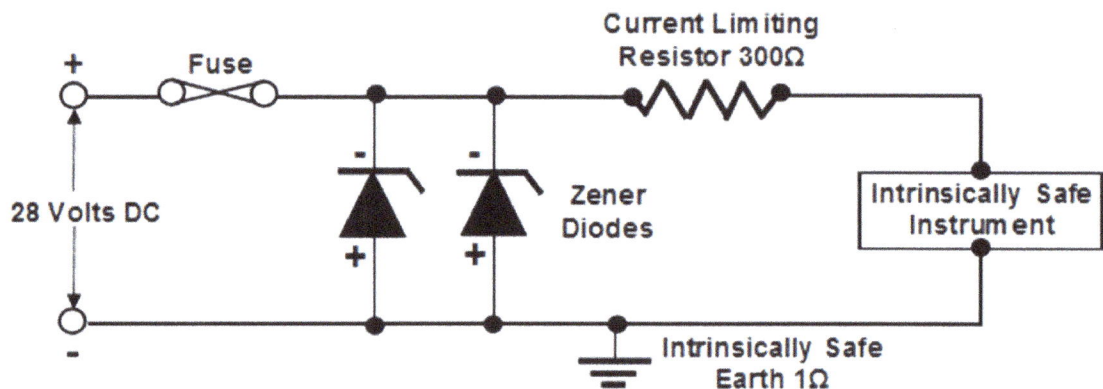

The supply (28V) positive (+) is connected to the Zener Diode cathode (-) which would mean of course that it is connected in reverse bias.

Let us suppose that the Zener Diodes in this barrier had a safety factor of 1.5 which means that if we multiply 28 x 1.5 = 42 so the breakdown voltage of the barrier would be beyond 42 volts, so provided that the voltage remained between 28 volts and 42 volts the diodes would not pass down to earth.

If a higher voltage than 42 volts, which could cause danger to the equipment in the hazardous area, invaded the loop/circuit then because of the breakdown effect the Zener Diodes would pass down to earth and blow the fuse. This is called the Zener effect where the electric field in the depletion region literally drags the electrons from their valance in the depletion zone.

We can now say that with a Standard Diode it was a one-way valve, with the Zener Diode, in our analogy being in actual fact, a **Pressure Relief valve.** Many diodes use this breakdown voltage in the way that they operate their protection. Also see **Backward Diode & Avalanche Diode**.

Flyback Diode Description:

This Diode may have other names such as Freewheeling Diode, Snubber Diode, Commutating Diode, Clamp Diode or Catch Diode. It is used on circuits where there is a high inductive load such as a solenoid valve where the coil may have hundreds of turns.

We had a situation where relay contacts were getting very badly pitted and we came to the correct, obvious conclusion that it was, in fact, that as the relays were controlling solenoids it was the solenoid valve coils that were causing it. We fitted Diodes in parallel with the coils and stopped the pitting.

Obviously pitting will be worse if the relay contacts are opened regularly depending upon the duty and size of the solenoid they control.

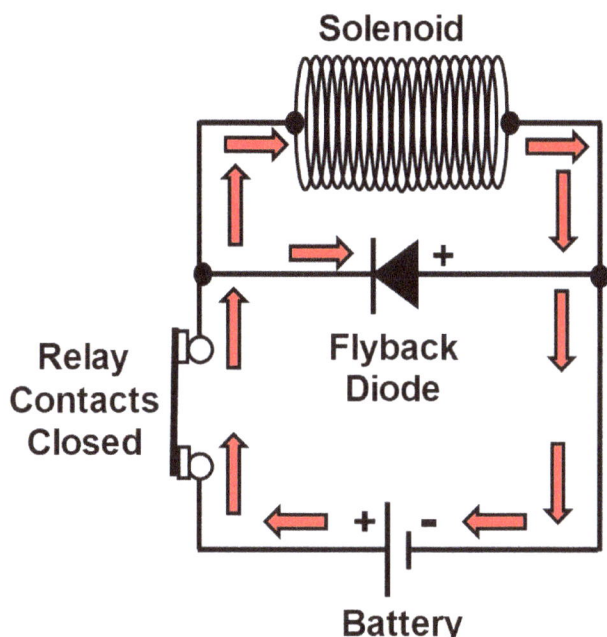

The diagram to left shows a simplified version of our circuit with the contacts of the relay operating a huge coil.

The Diode, as you can see, is connected in reverse bias to the power supply, meaning that the negative of the Diode is facing the positive of the supply.

Red arrows show current flow. So the current flows from the positive of the battery through the relay contacts, stops at the Diode because it is in reverse bias, through the solenoid and back to the battery.

Whilst in this position with the relay contacts closed there is no problem at all. The problems occur when the relay contacts open.

When the contacts are opened, which is a fast action, there is a phenomena called 'electromagnetic induction' within the solenoid coils as the magnetic field collapses, which causes a short high voltage spike.

This whole process is called 'Flyback' and will severely pit the contacts of the relay or switch unit.

As above, with power on, the Flyback Diode will be in reverse bias so will have no effect on the circuit.

When the relay contacts are opened a loop is formed between the Diode, which now changes to forward bias, and the solenoid coil dissipating the voltage away from the relay contacts. Shown with the blue arrows.

The result being that there is no pitting of the relay. contacts.

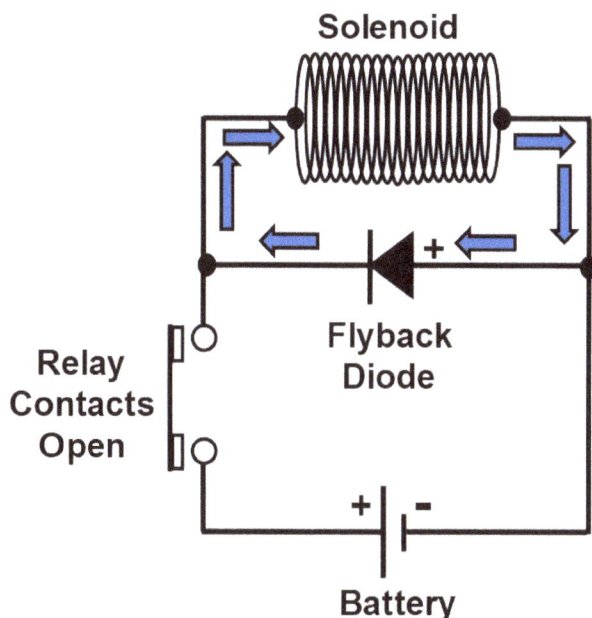

When the contacts open, which, as stated above, is a fairly fast action the voltage, caused by the collapsing magnetic field, momentarily jumps the air gap between the relay contacts and the circuit. This could be thousands of volts, similar to the principal of an electric fence control. If the contacts are operating fairly regularly you can see that there will be problems with severe pitting.

Flyback Diodes are not necessarily a special Diode in their own right. Providing that the Diode can cope with the voltage and current there is no problem, but manufacturer's advice is always the best option. They are only used where the inductor is controlled by relay contacts or a switch.

Anodes & Cathodes on Diodes:

When installing Diodes or Rectifiers into a circuit how do I know which way round to connect the Diode or Rectifier, i.e. forward (standard) or reverse bias? Below are six common Diodes and the diagram shows how the cathode is marked in each case:

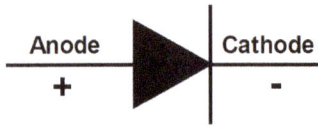

To the left is a Standard Diode symbol. In forward bias, the current flow is in the direction of the arrowhead and cannot return back against the arrow so how do Diode designs fit here? The Cathode is the blocking line.

To the left is a Standard Diode design. The Cathode of this Diode is the white band to the right hand side which matches the line after the arrowhead on the symbol.

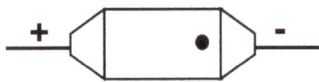

To the left is a white bodied Diode design. The Cathode of this Diode is the dot to the right hand side.

To the left is a glass Diode design. The Cathode of this Diode is the black area to the right hand side. Glass Diodes, I might add, will not always be white. Another colour is orange as in the case of a Zener Diode below.

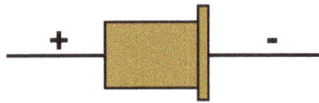

To the left is what is called a 'top hat' design. The Cathode of this Diode is the rim to the right hand side. These are a metal Diode usually gold or silver in colour.

To the left is a glass design Zener Diode. The Cathode of this Diode is the black band to the right hand side. These are usually connected in reverse bias as they are a 'Protection' Diode.

No matter what design of Diode you choose, the Cathode will be clearly marked in each case so you should not go wrong. Even multi-plate rectifiers will be marked in some way as to their Cathode.

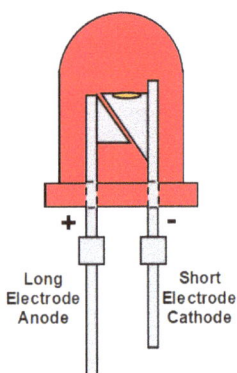

One diode that everyone takes for granted is the LED: Light Emitting Diode. It must be remembered that this unit is still a Diode and hence will have an Anode (+) and a Cathode (-) and is polarity conscious.

The LED will only light when the voltage is in forward bias so connect it in reverse bias and like any other Diode it will not work as it will not let the current flow in reverse.

Again like any other Diode if the current is high enough in reverse bias the Diode will break down and destroy itself.

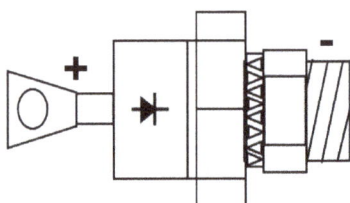

On the left is what is called a High Powered Metal Shell Diode. These diodes can go up to fairly high voltage, around 1000 Volts at high current of over 300 Amps. They can be Avalanche Diodes. They will have a mark showing the Anode and Cathode on the side as in the diagram.

Diode/Rectifier/Thyristor Symbols:

Below are some examples of Diode Symbols. Sometimes there may be three different symbols for one particular Diode but I have only shown one.

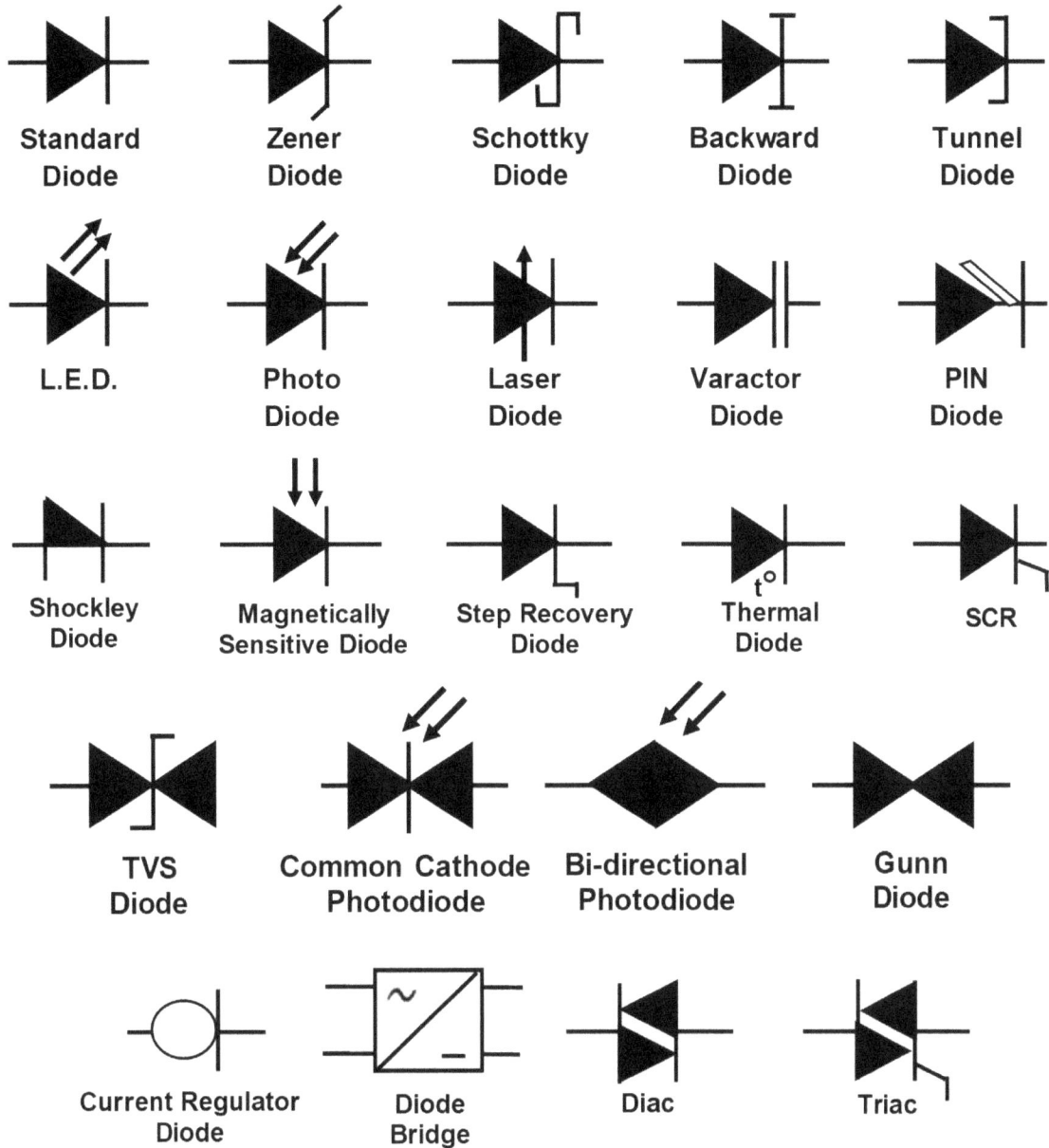

Standard Diode	**Zener Diode**	**Schottky Diode**	**Backward Diode**	**Tunnel Diode**
L.E.D.	**Photo Diode**	**Laser Diode**	**Varactor Diode**	**PIN Diode**
Shockley Diode	**Magnetically Sensitive Diode**	**Step Recovery Diode**	**Thermal Diode**	**SCR**
TVS Diode	**Common Cathode Photodiode**	**Bi-directional Photodiode**	**Gunn Diode**	
Current Regulator Diode	**Diode Bridge**	**Diac**	**Triac**	

Choosing the correct Component:

1) What do you want the Component to do? What is going to be its purpose in the circuit?

2) Is the Component going to be connected in forward or reverse bias? (Diodes).

3) What sort of current is the Component going to be dealing with?

4) What sort of voltage is the Component going to be dealing with? Are there any transient voltages?

5) What physical size of Component do you want, to be able to physically fit into the circuit?

6) What will be the operating temperature of the Component? This could have an effect on the current.

7) Is it a replacement? Does the component specs match what is already there?

These are the basic factors. There are many more characteristics to be looked into.

Testing a Diode:

Sometimes tests carried out on Diodes can seem to be ok, but unless approved methods are used then false readings may cause you to scrap perfectly good components.

The first thing to remember is to **ISOLATE THE POWER SUPPLY** before attempting to remove the Diode to be tested from the circuit, mainly for your own safety, but also for other components which may be damaged.

How to test a diode using a multi-meter on RESISTANCE:

So just to recap, we are going to test a conventional Diode using a Multi-meter. Looking at the selector you would think it was a case of just selecting Ω ohms and putting the probes on the Diode and it will read one way (forward bias) and not the other (reverse bias). On some very basic multi-meters that is more or less exactly what you would do as it would not offer you any more choices.

To test via the **RESISTANCE METHOD** firstly remove the diode from the circuit, or at least one end as leaving it in circuit could give false results.

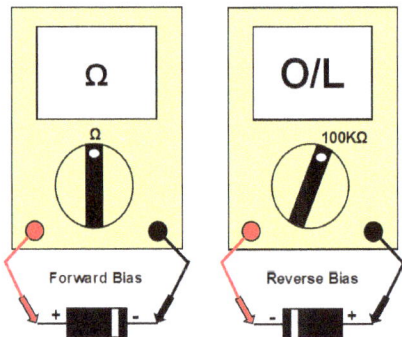

Using pure resistance on the standard multi-meter will determine if the Diode will pass current one way but not the other by connecting in forward and reverse bias as in the diagram on the left.

So in forward bias you would be looking at a lower resistance reading, but in reverse bias there will be a very high resistance approaching infinity, where the meter might read O/L overload. This may indicate a good Diode. This resistance method is not the best way to test Diodes as explained below.

In forward bias the reading should be around 1-10MΩ.

Using the resistance method above is sometimes not suitable. The multi-meter reads because when you put the probes onto the Diode you are actually putting on a voltage across a P & N junction which is the value of the battery in the instrument.

Sometimes this voltage may not be high enough to overcome the value of the Diode's forward potential. You could obtain readings that were not suitable on perfectly good Diodes and, in extreme cases, the reading could differ from instrument to instrument.

How to test a diode using a multi-meter on DIODE SETTING:

More expensive multi-meters have another setting for testing Diodes other than resistance and this setting has a symbol that looks like a Diode. This uses volt drop and not resistance so the reading here would be in volts not ohms.

As you can see in the diagram to the right, the multi-meter is set for Diode testing and this setting should be used for any semiconductor unit. The symbol on the test instrument is that of a Diode.

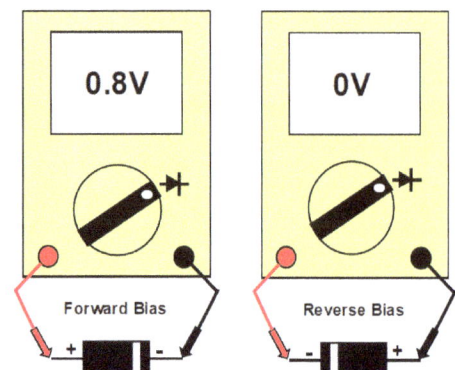

The test instrument applies a voltage across the Diode and measures the voltage drop **NOT THE RESISTANCE.**

The Silicon Diode may read a higher volt drop in forward bias than, say, a Germanium Diode. (Silicon around 0.8V Germanium around 0.3V.) 0V or O/L no circuit.

I have compiled a list of Diodes, Rectifiers and Thyristors. I do not think that anyone has any notion of exactly how many devices there are. You may even know of devices that are not on my list! All are in Alphabetical order.

List of Diodes: The following list of over 160+ different Diodes shows just how many types of Diode have been found and there may actually be more!

1 - Automotive Diodes: Used in the automobile industry.

2 - Avalanche Diode: Protection Diode used in reverse bias. This is similar to the Zener Diode. Full description earlier.

3 - Avalanche Photodiode (APD): Used for Optical communication and accurate distance measurement. Suitable for high frequency. Full description earlier.

4 - Avalanche Transit Time Diode (ATT): Group name for IMPATT: Impact Ionisation Avalanche Transit-Time Diodes, BARITT: Barrier Injection Transit Time Diode & TRAPATT: Trapped Plasma Avalanche Triggered Transit Diode. See each for information.

5 - Axial Lead Diode: Any diode where the leads form an 'axis' through the Diode. Full description earlier.

6 - Axial Rectifier Diode: Similar to a Standard Diode with higher current carrying capacity. Used in case of reverse connection.

7 - Back Diode: Another name for the Backward Diode. (Next)

8 - Backward Diode: Protection Diode so connected in reverse bias. This is similar to the Zener Diode.

9 - BARITT Diode: Barrier Injection Transit Time Diode. **See** Avalanche Transit Time Diode (ATT). Used in microwave signal generation.

10 - Bi-directional Diode: Another name for this Diode is DIAC: Diode for Alternating Current or Bi-Polar Suppressor Diode. DIACs are used to trigger Thyristors.

11 - Bi-directional (TVS) Diode: Transient Voltage Suppressor Diode. Protection diode for protecting integrated Circuits.

12 - Bi-Polar Suppression Diode: See Bi-Directional Diode. Protects against transient voltages. Used in Telecommunications.

13 - Blocking Diode: Used with Solar Panels for their protection. As a Diode should, it prevents current flowing back to the solar cells.

14 - Blue Laser Diode: As the title a Laser Diode. Typically constructed from Aluminium Gallium Nitride.

15 - Body Diode: Also Bulk Effect Diode & also part of the circuit of a MOSFET: Metal Oxide Semiconductor Field Effect Transistor. May be called 'Internal' Body Diode or Parasitic Diode.

16 - Breakdown Diode: Protection Diode. Connected in reverse bias & may also be called a Zener Diode or Voltage Regulator Diode.

17 - Briggs Diode: This is a Trade Name.

18 - Bulk Effect Diode: Also Body Diode & also part of the circuit of a MOSFET: Metal Oxide Semiconductor Field Effect Transistor. May be called 'Internal' Body Diode.

19 - Bypass Diode: Used with solar panels.

20 - Capacitive Diode: It is a Diode, but operates like a Capacitor with its P & N regions moving together and apart depending upon voltage and forward and reverse bias.

21 - Catch Diode: For the protection of relay contacts with inductive loads. Can be called a Flyback Diode.

22 - Charge Storage Diode: Is called a Diode, but works more like a Solid State Switch than a Diode.

23 - Clamping Diode: This Diode 'clamps' the voltage whereas a Zener Diode waits for a rise in voltage over a certain value.

24 - Clipping Diode: Bit like a Thyristor where it 'clips' the voltage at a certain value.

25 - Common Anode Diode: CAN: Control Area Network & ESD: Electrostatic Discharge.

26 - Common Cathode Diode: CAN: Control Area Network & ESD: Electrostatic Discharge.

27 - Common Cathode Laser Diode: Used in Printers, Bar Code Readers etc.

28 - Commutating Diode: See Flyback Diode, Kickback Diode & Snubber Diode.

29 - Constant Current Diode (CCD): Also known as a CLD: Current Limiting Diode & CRD: Current Regulating Diode.

30 - Crystal Diode: Used in the old Crystal Radios.

31 - Current Limiting Diode (CLD): Also known as a CCD: Constant Current Diode also CRD: Current Regulating Diode. Used in battery chargers & timing circuits.

32 - Current Regulating Diode (CRD): Protection Diode for analogue circuits. Also known as a CLD: Current Limiting Diode or CCD: Constant Current Diode.

33 - Data Diode: These control Data rather than current.

34 - Detector Diode: Sometimes called a Germanium Point Contact Detector Diode.

35 - DIAC: Diode for Alternating Current. Also see Bi-Directional Diode. Used to trigger Thyristors.

36 - Diode Bridge: Full wave rectification. Description earlier. Another name is a Graetz Diode Bridge.

37 - Discrete Diode: Complete zero resistance to current in forward bias.

38 - Double Base Diode: Although it is called a Diode, it really is not a Diode as such, it can be called a Unijunction Transistor.

39 - Double Colour Light Emitting Diode: See LED: Light Emitting Diode.

40 - DOVETT Diode: Double Velocity Transit Time Diode. Related to the BARITT Diode: Barrier Injection Transit Time Diode.

41 - Dynistor Diode: Another name for this item is a Diode Thyristor. A smaller version of the Shockley Diode with smaller signals. Now obsolete & replaced by the Unidirectional Thyristor Breakover Diode.

42 - Edge Emitting Diode: Similar to, but slightly different from the LED: Light Emitting Diode.

43 - Edge Emitting Laser Diode: These are solid state, semiconductor Laser Diodes.

44 - Esaki Diode: Another name is a Tunnel Diode.

45 - ESD: Electrostatic Protection Diode. Also TVS: Transient Voltage Suppression Diode. Very similar to the Zener Diode.

46 - Fast Recovery Diodes: FRD Diodes are what is known as a FRTD: Fast Recovery Time Diode. Used in digital communications.

47 - Fast Switching Rectifier Diode: See Shockley Diode and Switching Diode.

48 - Flyback Diode: For the protection of relay contacts with inductive loads. Can be called a Catch Diode, Freewheeling Diode Commutating Diode or Snubber Diode.

49 - Forward Reference Diode: Also called a 'Stabistor'. This is a Silicon semiconductor Diode capable of a 'stable' forward low voltage, hence its other name of 'Stabistor'.

50 - Four Layered Diode: Sometimes called a PNPN Diode or Shockley Diode.

51 - Freewheeling Diode: For the protection of relay contacts with inductive loads. Can be called a Catch Diode, Flyback Diode Commutating Diode or Snubber Diode.

52 - Full Wave Controller Rectifier Diode: See Centre Tapped Full Wave Rectifier and Diode Bridge.

53 - Gallium Arsenide Diode: Gallium Arsenide (GaAs) is the semiconductor used in this Diode as against a Silicon semiconductor Diode.

54 - Germanium Diode: Germanium semiconductor was used in older Standard Diodes, but has made a comeback in some modern Diodes because of its low volt drop.

55 - Gold Doped Diode: Expensive Diodes, because Gold or Platinum is used for doping,

56 - Glass Passivated Diode: In the preparation/development stage at the moment.

57 - Graetz Diode Bridge: See Diode Bridge. Full wave rectification.

58 - Gunn Diode: Also called a TED: Transferred Electron Device. Used in Speed Guns etc.

59 - Gunn Diode Oscillator: Also known as a Gunn Oscillator. Generate microwave frequencies.

60 - Gunn Oscillator: See Gun Diode Oscillator above.

61 - Half Wave Controller Rectifier Diode: A Standard Diode inserted into an AC circuit will in fact produce half wave.

62 - High Backpressure Switching Diode: This semiconductor Diode action is more of a switch than a Diode as the name might suggest. Usually switches very small currents in the mA range and low voltage. See Switching Diode & Shockley diode.

63 - High Current Silicon Rectifier: These handle high power and are more of a Thyristor but employ the function of a Diode. Also see Rectifier section.

64 - High Temperature Diodes: Items such as this make excellent Thermometers as the voltage increases or decreases depending on the temperature of the unit. See Low Temperature Diode.

65 - High Speed Switching Diode: See Switching Diode.

66 - High Voltage Diodes: These are standard Diodes, but are exactly what the title states; they take High Voltage.

67 - Ideal Diode: This is exactly what the title says; it acts as an ideal conductor when a voltage is put onto it with a forward bias, and an ideal insulator when the voltage is put onto it in reverse bias. No threshold or start up voltage, it starts conducting immediately. The current will flow as if you had just closed a switch!

68 - IMPATT Diode: Impact **I**onisation **A**valanche **T**ransit **T**ime Diodes. Produces 'intelligent' radiation.

69 - Infrared Laser Diode: Ideal for machining applications.

70 - Infrared Receiver Diode: Works similar to a Photodiode & Optodiode.

71 - Injection Laser Diode (ILD): Very similar to an LED: Light Emitting Diode.

72 - In-Plane Laser Diode: See Injection Laser Diode CD & DVD players.

73 - Internal Body Diode: Another name for a Body Diode.

74 - Intrinsic Body Diode: See MOSFET Body Diode: Metal Oxide Semiconductor Field Effect Transistor. Used in converters.

75 - Large Signal Diode: Handles a fairly high current as against Small Signal Diodes.

76 - Laser Diode: A semiconductor Laser Diode uses its P & N junction to form a laser type light.

77 - Linear Diode Array (LDA): This diode detects light using several photodiodes in lines.

78 - Light Emitting Diode (LED): Includes a Double Colour Light Emitting Diode. Full description earlier.

79 - Low Power Switching Diode: See Switching Diode.

80 - Low Temperature Diodes: Items such as this make excellent thermometers as the voltage increases or decreases depending on the temperature of the unit. See High Temperature Diode.

81 - Magnetic Diode: This is fairly new technology. All I can say is that the Diode controls current using a magnetic field instead of semiconductors.

82 - Magnetically Sensitive Diode (MSD): Sometimes called a Magneto Diode; as the name suggests this diode is very sensitive to magnetic fields.

83 - Magneto Diode: See Magnetically Sensitive Diode above.

84 - Microwave Diode: There are many Diodes in this section that operate in the Microwave Frequency band for example BARITT Diodes: Barrier Injection Transit Time Diode, Avalanche Diodes, PIN Diodes, Varactor Diodes etc.

85 - MITATT Diodes: Mixed Tunnelling and Avalanche Transit Time Diode. Low noise output.

86 - MOSFET Body Diodes: These are usually Standard Diodes that are part of a circuit rather than individual Diodes. They lie between the Drain and Source of a MOSFET: Metal Oxide Semiconductor Field Effect Transistor.

87 - MOV: Metal Oxide Varistor. Many uses as Protection Diodes.

88 - Multiplier Diode: This really is a circuit with Diodes and many capacitors that can, as per the title, multiply the DC output voltage to much more than the AC input voltage.

89 - Noise Diode: One of the biggest noise generation in electronic items is the movement of electrons in electronic equipment such as certain Diodes.

90 - NPN Triode: Used to amplify weak signals. See also PNP Triode.

91 - Offset Diode: Used together with a battery to offset voltage and current in an ideal Diode.

92 - Optodiode: Another name is the Photodiode. Used for things like flame detection.

93 - Ordinary Switching Diode: See Switching Diode.

94 - Parasitic Diode: Another name for a Body Diode.

95 - Parasitic Capacitance Diode: Protection Diode. Passes high frequency signals and does not block them.

96 - Peltier Diode: Another name is a Thermal Diode. Used on small refrigeration devices.

97 - Photodiode: See Optodiode, Photovoltaic Diode & Infrared Receiver Diode.

98 - Photo-Interrupter Diode: I have called this a diode but is more of a switch. Does a very similar task to a Magnetic Mobrey Switch except this one works on light.

99 - Photovoltaic Photodiode: See Photodiode.

100 - Pico Amp Diodes: Very low leakage Diodes.

101 - PIN Diode: it is a cross between an Avalanche Photodiode and a Capacitive Diode.

102 - PIN Photodiode: These are better than a PN Photodiode when the light level is not so bright.

103 - PN Junction Diode: See Standard Diode.

104 - PNP Triode: Used to amplify weak signals. See NPN Triode.

105 - PN Photodiode: See PIN Photodiode above.

106 - PNPN Diode: See Four Layered Diode.

107 - Point Contact Diode: The 'Cats Whisker' used in early radios, well this Diode is a variation on that theme.

108 - Power Diode: Although they operate the same as Standard Diodes they are big bulky metal units.

109 - Precision Diode: Can be used as a Half or Full Wave Rectifier or Clipper Diode.

110 - QPL Diode: Stands for Qualified Products List, which means that these particular Diodes, of whatever type, appear on an approved list for, say, organisations such as the military.

111 - Real Diode: See Standard Diode.

112 - Rectifier Diode: See Super Barrier Diodes.

113 - Red Laser Diode: See Injection Laser Diode.

114 - Reverse Polarity Protection Diode: Stops equipment from being ruined due to connecting in reverse polarity.

115 - Selenium Diode: its correct name is the Selenium Rectifier Diode. Handles high current and can be replaced by Silicon these days.

116 - Schottky Diode: Fast switching action with low forward volt drop which makes them attractive. See Super Barrier Diodes & Rectifier Diodes.

117 - Shockley Diode: Not to get mixed up with the Schottky Diode, like the SIDAC: Silicon Diode for Alternating Current. Is again more of a 4 layered fast switch than a Diode.

118 - SiC Diodes: Silicon Carbide. Used in Inverters, motor drives and UPS.

119 - SIDAC Diode: Silicon Diode for Alternating Current. This is really a semiconductor, Bidirectional AC Switch more than a Diode, a bit like a Transistor.

120 - Signal Diode: same as Switching Diode. Used a lot in the Computers.

121 - Silicon Diode: Many Diodes are Silicon semiconductor based units.

122 - Silicon Voltage Switching Diode: See Switching Diode. Could be a SIDAC.

123 - Single Photon Avalanche Diode (SPAD): Bio-photonics.

124 - Small Signal Diodes: Small current high frequency applications.

125 - Snap off Diode: See Step Recovery Diode.

126 - Snubber Diode: See Flyback Diode.

127 - SPAD: See Single Photon Avalanche Diode.

128 - Stabistor: Also called Forward Reference Diode.

129 - Standard Diode: See Standard Diode at the beginning.

130 - Steady Regulator Voltage Bi-polar Diode: Very similar to the Zener Diode.

131 - Step Recovery Diode: This, as most of the Diodes I mention, is a Semiconductor Diode and it works more like a Switch/Capacitor than a Diode.

132 - Super Barrier Diode: These Diodes are known as Rectifier Diodes and are the next generation of Schottky Diodes.

133 - Suppressor Diode: Known more as Transient Suppressor Diodes and can be unidirectional or bidirectional. These diodes are usually part of a circuit that diverts transient spikes away from delicate electronic circuits. See Flyback Diode.

134 - Surface Emitting Laser Diode: This unit emits light perpendicular to the active region.

135 - Switching Diode: This semiconductor Diode action is more of a switch than a Diode as the name might suggest. Usually switches very small currents in the mA range and low voltage. See Fast Switching Diode & Shockley Diode. Could be a SIDAC: Silicon Diode for Alternating Current.

136 - Temperature Sensing Diode: The resistance of the Diode changes with temperature.

137 - Temperature Sensitive Diode: Used to measure temperature.

138 - Thermal Diode: Another name is the Peltier Diode. Used on small refrigeration devices.

139 - Three Phase Diode Converter: Can be a Three Phase Bridge Rectifier.

140 - Through Hole Diodes: Used in Inverters.

141 - Thyrector Diode: Other names for this Diode are TVS: Transient Voltage Suppression Diode and Transil Diode.

142 - Top Hat Diode: Refers to the shape of the Diode resembling a top hat.

143 - Transient Suppressor Diodes: Do a very similar job to a Zener Diode in protecting delicate systems from high transient voltages. See Zener Diode.

144 - Transil Diode: A more common name for this Diode might be a TVS: Transient Voltage Suppressor or Thyrector Diode. This is a protection Diode not too dissimilar from the Zener Diode.

145 - TRAPATT Diode: Stands for Trapped Plasma Avalanche Triggered Transit Diode. Is a member of the IMPATT Diode Impact Ionisation Avalanche Transit Time Diodes family in the way that it operates. Capable of switching higher currents much faster than other Diodes.

146 - Trisil Diode: This is a protection diode such as a Zener Diode.

147 - Tube Diode: See Glass tube Vacuum Diode. Quite old.

148 - Tuner Diodes: Another name for Varactor Diodes. Used in voltage controlled oscillators, harmonic generators etc.

149 - Tunnel Diode: Protection Diode so connected in reverse bias. Also see Backward Diode & ESAKI Diode.

150 - TVS Diode: See Transil Diode and Thyrector Diode. Also see ESD Protection Diode.

151 - Ultra-Fast Recovery Rectifier Diode (UFRRD): are what is known as a UFRT: Ultra Fast Recovery Time Diode. Used in digital communications.

152 - Ultra-Fast (High Speed) Switching Diode: See Switching Diode.

153 - Unidirectional Thyristor Breakover Diode: Replaced the Dynistor.

154 - Vacuum Diode: This of course is a Glass Tube Diode of the past. Works in exactly the same way as a conventional diode i.e. only allowing current to flow in one direction.

155 - Varactor Diode: See Capacitive Diode, Tuning Diodes & Step Recovery Diode. Used in voltage controlled oscillators, harmonic generators etc.

156 - Variable Diode: Variable Diodes can be for example Variable Capacitance Diode or Variable Zener Diodes.

157 - Varicap Diode: Variable Capacitance Diode. Used in voltage controlled oscillators.

158 - Varistor Diode: This electric component name is made up from Variable Resistor.

159 - Variable Capacitor Diode: Used for tuning circuits.

160 - Variable Zener Diode: Acts like a normal Zener Diode, but can be adjusted.

161 - Voltage Regulator Diode: See Zener Diode & Breakdown Diode.

162 - Voltage Adjustment Diode: This is more of a circuit than a single diode!

163 - Wide Bandgap Diode: This refers to the semiconductor P & N properties rather than being an actual Diode. It allows items such as Diodes to operate at much higher voltages.

164 - Xray Diode: These are part of very complex units and used in the measurement of Xray Flux.

165 - Zener Diode: May be called **Voltage Regulator Diodes or Breakdown Diodes**. Used to protect delicate electronic components especially Intrinsically Safe Circuitry in Hazardous Areas.

List of Rectifiers:

The difference between a Diode and a Rectifier is that a Rectifier could be a full circuit which may include a diode. I have listed 31 Rectifiers – there may be more! Rectifiers can be controlled or uncontrolled, three phase or single phase, half wave or full wave. All are in Alphabetical Order.

1 - Active Rectifier: Converts AC to DC without power loss or heat.

2 - Three Phase Bridge Rectifier: Uses six Diodes to form a three-phase bridge, again from a transformer 'starred' secondary.

3 - Centre Tapped Full Wave Rectifier: Utilises two Diodes in the Rectifier Circuit.

4 - Controlled Bridge Rectifier. Does not use uncontrolled Diodes. Uses MOSFETs: Metal Oxide Semiconductor Field Effect Transistors or SCRs: Silicon Controlled Rectifiers.

5 - Controlled Centre Tap Rectifier: Produces full wave rectification.

6 - Copper Oxide Rectifier: Used in the 1920s, Copper Oxide Semiconductor Rectifiers.

7 - High Current Silicon Rectifier: Current ratings can go as high as 100A+.

8 - High Frequency Rectifier: High frequency as per the title used in telecommunications.

9 - High Voltage Rectifiers: As in the title can handle high voltage. Can be around 200kV. Can be Leaded Silicon or standard.

10 - Negative Half Wave Rectifier: The Rectifier is connected reverse bias.

11 - Plating Rectifier: Used in electroplating.

12 - Positive Half Wave Rectifier: The Rectifier is connected forward bias.

13 - Rotating Rectifier: This would be on the exciter of a motor to revert the AC to DC.

14 - Selenium Rectifier: These really have been superseded by the Silicon Diode.

15 - Silicon Controlled Rectifier (SCR): Sometimes called semiconductor Rectifiers. This really is another name for a Thyristor which is described later.

16 - Single Phase Controlled Rectifier: Uses Thyristors instead of Diodes. Can be used for Motor Control.

17 - Single Phase Half Wave Controlled Rectifier. Really this is a Thyristor circuit.

18 - Single Phase Full Wave Controlled Rectifier. A Bridge Rectifier with Thyristors instead of Diodes.

19 - Single Phase Rectifier: Usually used for UPS: Uninterruptible Power Supply. Can be Diodes.

20 - Single Phase Half Controlled Rectifier: Uses Thyristors instead of Diodes.

21 - Single Phase Uncontrolled Rectifier: Usually achieved with Diodes.

22 - Split Supply Rectifier: Done with a centre tap transformer and a Diode Bridge.

23 - Synchronous Rectifier: Similar to a Diode passing current in one direction. Can use a MOSFET: Metal Oxide Semiconductor Field Effect Transistors.

24 - Three Phase Controlled Rectifier: This Rectifier uses Thyristors for controlled Rectification as against Diodes for Uncontrolled Rectification.

25 - Three Phase Full Wave Rectifier. Usually uses six Diodes in the double secondary of a three-phase transformer.

26 - Three Phase Half Controlled Rectifier: Three Diodes are connected into the secondary of a three phases of a star connected transformer.

27 - Three Phase Half Wave Rectifier: Three Diodes connected in star.

28 - Three Phase Rectifier: Uses Diodes in each of the three phases. Usually from the 'starred' secondary of a three-phase transformer.

29 - Three Phase Uncontrolled Rectifier: This Rectifier uses Diodes to achieve its objective which are of course uncontrolled.

30 - Vibrating Rectifier: These are a device with a resonant 'reed' vibrated by an AC magnetic field.

31 - Voltage Multiplying Rectifier: Can double, triple and quadruple a voltage. Using multiple doublers, this device can quadruple a DC voltage.

The Thyristor Section:

Many experts may disagree with me but I like to describe a Thyristor as a Diode with a Gate, which I personally think is a description that people can understand.

They can appear to be very similar to a Transistor, which in all fairness they are. They both require power on the gate to work. The main difference between the two is that when the power is put onto the gate, most Thyristors latch whereas Transisters do not.

In fact the word Thyristor is formed from two other names of different equipment i.e. **Thyr**atron and Trans**istor**; a Thyratron being a gas filled Triode. In the past, before the Thyristor came along, the Thyratron was used for switching where the load was fairly large. The Thyristor is much faster.

Because Silicon is the primary material in the manufacture, there is another name for a Thyristor and that is an SCR (**S**ilicon **C**ontrolled **R**ectifier).

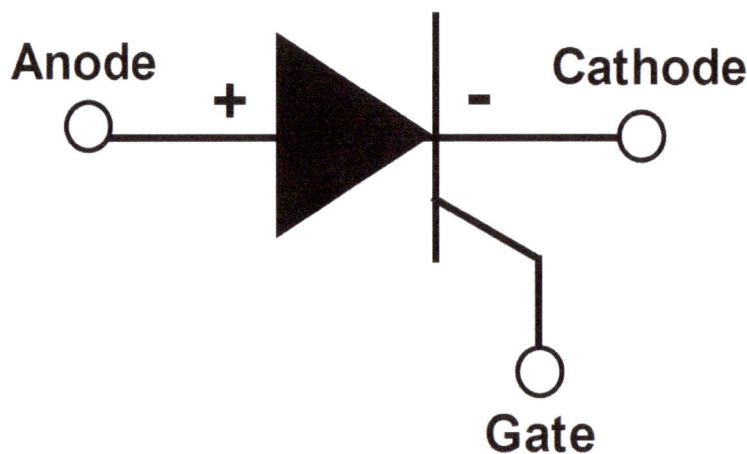

On our factory complex one of the main pieces of equipment that required extensive use of Thyristors was an Inverter.

Back in the 1990s Inverters were used for Emergency Lighting and Instrument Inverters and I can tell you that they were extremely noisy. They ran constantly and the huge battery banks were checked regularly.

Let us in the next few pages look at how the Thyristor operates and how many different types of Thyristor are available to us:

NOTE:

ON ALL ATEX EQUIPMENT FOR HAZARDOUS AREAS SUCH AS ZONES, REMOVAL/CHANGING OF THYRISTORS FROM A MANUFACTURERS' PRINTED CIRCUIT BOARD CAN ONLY BE CARRIED OUT WITH MANUFACTURER APPROVAL IN WRITING.

FAILURE TO COMPLY WITH THIS MAY BE CLASSED AS AN UNAUTHORISED MODIFICATION AND MAY RENDER ANY CERTIFICATION DOCUMENTATION NULL AND VOID.

ALSO THE DESIGN, MANUFACTURER, TYPE ETC., OF THE THYRISTOR COMPONENTS MUST NOT BE CHANGED IN ANY WAY.

Thyristor Description:

Thyristors are equal to an SCR, **Si**licon **C**ontrolled **R**ectifier, which will do the same job. They are made up of four layered regions with three PN Junctions. Really we can say that they are a Diode with a gate.

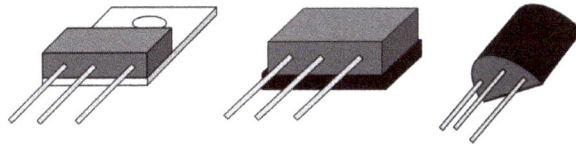

Standard Thyristors are types of very rapid, usually three electrode, switching devices, not too dissimilar from Diodes, that can be used in many modern day appliances such as laptops, televisions etc. Very similar to a higher powered Transistor and we can use them in similar ways.

Thyristors tend to carry much higher currents than a Transistor and there are instances where we want the device to remain **ON** and 'latch' and will only turn off when the Anode current drops to zero. On a Transistor, removing the gate current would switch it **OFF**.

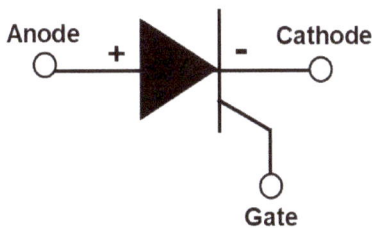

The Thyristor can be symbolised by the diagram to the left which shows it as a three terminal device having namely an Anode (+), Cathode (-) and Gate. The Thyristor, like a Transistor, would require voltage on the Gate before it will switch. Older industrial inverters would have large Thyristors in them to control their voltages and in the past these would be extremely noisy.

Up to now we have a Silicon Diode type unit that requires a Gate to be opened before it will allow current to flow. Let us now look at the above diagram in terms of a P & N three terminal diagram.

The P & N Diagram to the right shows that there are four layers & three junctions. There is an Anode and Cathode at either end with a switching Gate coming from the region 3 positive. So the current flow will be from left to right. Let us look at this from another angle; what we have is two Transistors back to back in series. The regions 1, 2 and into 3 are one Transistor, regions 4, 3 and 2 being the other. Let us move to the diagram below.

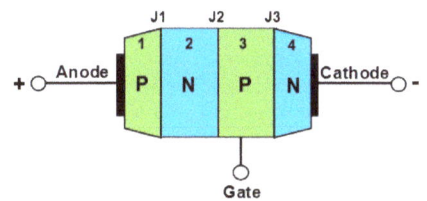

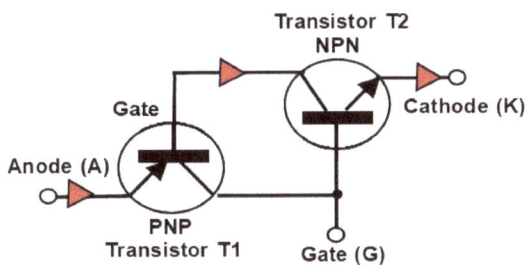

The diagram to the left now shows our two Transistors making up our Thyristor. As you can see we have Transistor 1 which is PNP and Transistor 2 which is NPN. Our three terminals are Anode (A), Cathode (K) and switching Gate (G). The red arrows show the direction of current when the Thyristor is working. The main difference at this stage between a Transistor and a Thyristor is that the Gate needs only a pulse to turn it **ON** and it will remain latched **ON**. A Transistor requires power constantly on the Gate to keep it **ON**.

Let us say that the Thyristor is connected in forward bias i.e. positive of the battery to terminal 'A'. We now put the pulse on the Gate that we have talked about; this is called a **'trigger'** pulse and will cause the Thyristor to 'latch' **ON** and operate, allowing current flow. The Thyristor has now become a Diode.

So long as the Anode current is higher than the Gate current then the Thyristor will remain latched **ON**. This is called the **'holding'** current and the Thyristor is in forward conducting mode, and in our case above Transistor T1 feeds Transistor T2. With only one Transistor we would have to keep current on the Gate permanently as it would not latch.

Removal of the Gate current at this stage will **NOT** turn it **OFF**. So how do we turn it off? To turn the Thyristor **OFF** we must now drop the **'holding'** current at the Anode and because there is no Gate current the device will turn **OFF** which we call **'forward blocking'** mode. Reverse blocking is achieved when the unit is connected in reverse bias just the same as a Standard Diode.

1) A Thyristor, drawing symbols below, is very similar to a Diode, but with this device, as with a Transistor, there is a **'Gate'** which requires to be operated.

2) So, like a Diode, in most cases, a unidirectional, semiconductor device with P & N junctions although RCTs (**R**everse **C**onducting **T**hyristor) can be obtained which are bi-directional.

3) They can be used as a type of clipper switch to control voltage levels and used in equipment such as Inverters.

4) Thyristors 'latch' and require the Anode current to drop to '0' to turn it off. It will not turn off by switching off the Gate current. Below are three Thyristor Symbols, I have drawn them in this section as bottom left **'A'** (I.E.E.E.)

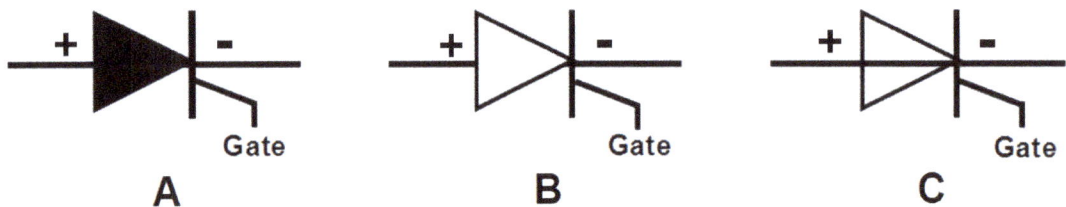

Thyristor Circuit:

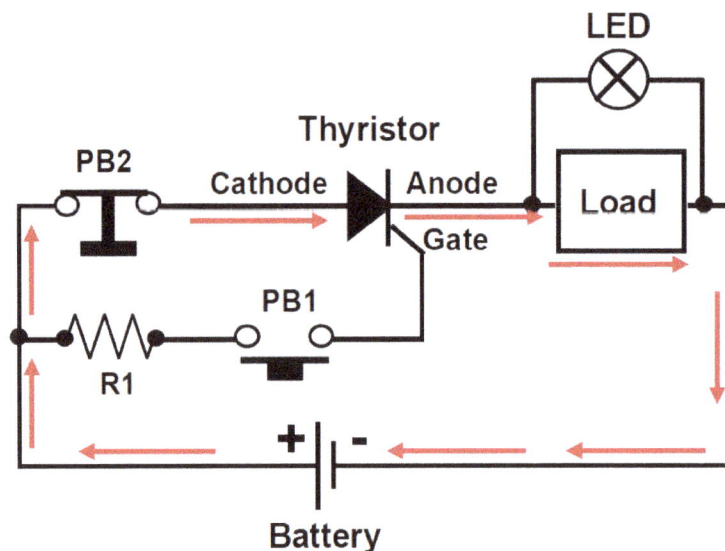

The diagram above shows a very simple Thyristor Circuit which could be done with just a switch but I have fitted a Thyristor for the purpose of my explanation. It shows the current flow **AFTER** the push switch PB1 has been operated.

In this diagram we have a Thyristor, a Battery, two Push Buttons and a Load/LED. The Thyristor is connected between the Battery and Load.

The difference here between the Thyristor and a Diode is that with a Diode, current would flow when the circuit is switched on, but with a Thyristor no current will flow until the Gate is energised.

Push Button PB1 is operated. So we push PB1 (N/O); now there is power on the Thyristor Gate, which allows current to flow in the direction of the red arrows. However releasing PB1 will not de-energise the Thyristor. We could call this a convenient 'Latch' Circuit.

The current at the Thyristor Cathode must drop to '0' before the current flow will cease and in our case that can be achieved by pressing PB2 (N/C).

States of a Thyristor:

Thyristors have different 'States', three to be exact, namely:

1) **Forward Blocking:** No current flowing. Because a Thyristor in effect is two Diodes in forward bias joined together hence the junction between is reverse bias, so without any current on the Gate to switch the Thyristor is blocking.

2) **Reverse Blocking:** No current flowing. Because a Thyristor in effect is two Diodes, but this time they are connected in reverse bias, again the Thyristor itself acting like a Diode in reverse is blocking.

3) **Forward Conducting:** This is the usual one. Two Transistors as per the diagram. The Anode is connected to positive which leaves the Cathode negative. When current is applied to the Gate the Transistors switch each other, from one to the other until fully saturated, then current flows.

How do they Latch?

Just follow the current numbers. We start off with Diagram A and the Thyristor is at rest. In Diagram B we then put currents 1 & 2 and switch the NPN Transistor 2 and we now have currents 3 & 4. In Diagram C switching the Gate of Transformer 1 we have currents 5 & 6. Now in Diagram D we remove current 1, **LATCHED**.

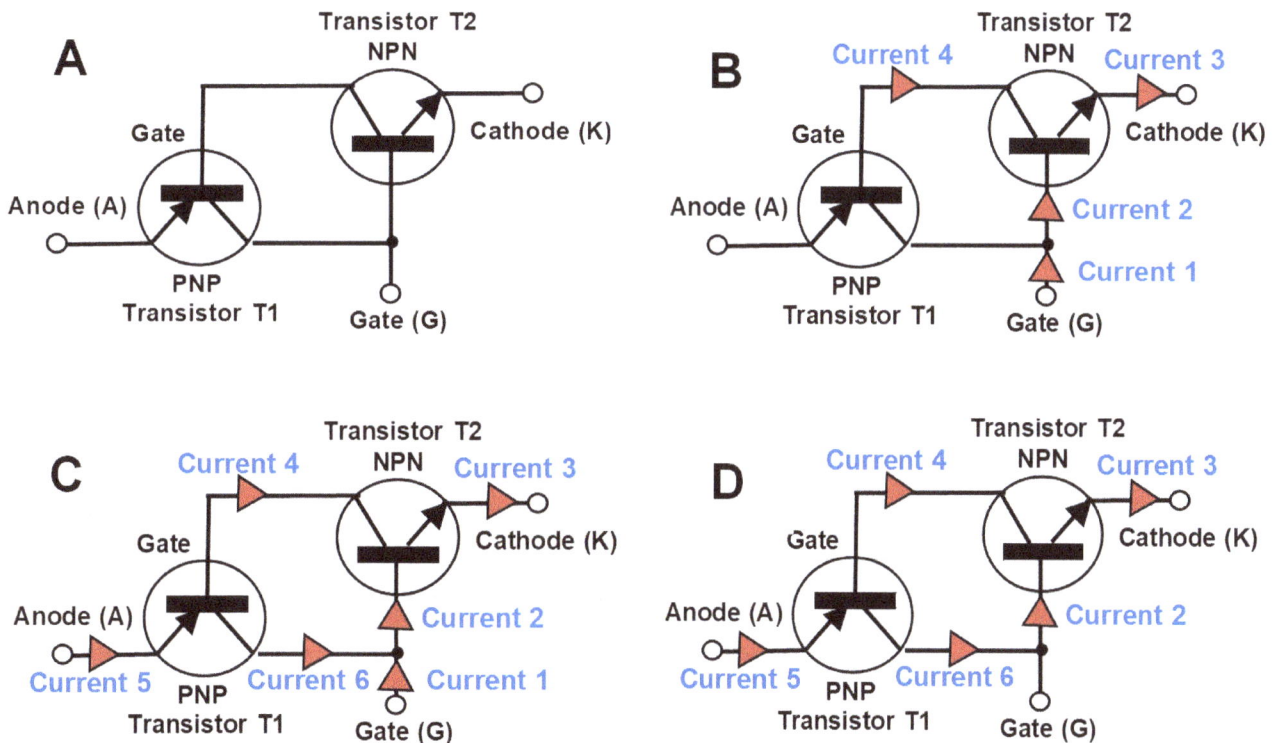

List of Thyristors:

Below is a list of Thyristors, remember these are Diodes with Gates. The actual name 'Thyristor' comes from being a cross between a **Thyra**tron and Trans**istor.** I have listed twenty-nine of them in Alphabetical order.

1 - Bi-directional Control Thyristor. (BCT)

2 - Bi-directional Triode Thyristor. (TRIAC): A three terminal switch used on AC so must be able to conduct in both directions.

3 - Darlistor. (Diode Thyristor): Used as a block for high voltages and current.

4 - Ditriac: This is a bi-directional.

5 - Emitter Turn-off Thyristor. (ETO): This uses MOSFET: Metal Oxide Semiconductor Field Effect Transistor and SCR: Silicon Controlled Diode/Rectifier. It also has two Gates as opposed to a standard Thyristor which only has one Gate. One Gate turns it on and the other Gate turns it off.

6 - Fast Switching Thyristor. (FST): Can be called an 'Inverter' Thyristor. Used on medium frequency equipment.

7 - Field Effect Transistor. (FET) Controlled Thyristor (FET-CTH): Uses a magnetic field to control the flow of current. Uses MOSFET: Metal Oxide Semiconductor Field Effect Transistor & SCR: Silicon Controlled Diode/Rectifier.

8 - Fourth Layer Schottky Thyristor. Thyristors are four layered devices i.e. PNPN with the Gate contact on the second 'P' layer.

9 - Gate Assisted Turn-off Thyristor. (GATT): The Gate bias acts so that a forward voltage does not appear at the Cathode using the P&N junction.

10 - Gate Turn-off Thyristor (GTO): Also see Turn off Conducting Thyristor. Uses a negative pulse on the Gate to turn it off.

11 - High Current Silicon Controlled Diode/Rectifier (SCR): More of a high current Thyristor.

12 - Integrated Gate Commutated Thyristor. (IGCT): Very closely related to the GTO: MOS Turn-off Thyristor.

13 - Light Activated Silicon Controlled Rectifier. (LASCR): Although this seems like a Rectifier it is in fact a Thyristor.

14 - MOS Controlled Thyristor. (MCT): Combines the properties of both a MOSFET: Metal Oxide Semiconductor Field Effect Transistor and SCR: Silicon Controlled Diode/Rectifier.

15 - MOS Turn-off Thyristor. (MTO): A high powered device which combines both MOSFET: Metal Oxide Semiconductor Field Effect Transistor and GTO: Gate Turn-off Thyristor.

16 - N Gate Thyristor: Connected to the 'N' region of a P & N so switching is done with a negative signal.

17 - P Gate Thyristor: Connected to the 'P' region of a P & N so switching is done with a positive signal.

18 - Photo Thyristor. Operated by light.

19 - Quadrac: Contains a DIAC: **D**iode for **A**lternating **C**urrent and a TRIAC: **Tri**ode **A**lternating **C**urrent.

20 - Reverse Blocking Triode Thyristor: In this case the blocking is in the reverse direction.

21 - Reverse Conducting Thyristor: Has an integrated reverse Diode so cannot block.

22 - Silicon Bilateral Switch (SBS): Working more of a Thyristor come Zener Diode.

23 - Silicon Controlled Diode/Rectifier (SCR): Another name is a Thyristor.

24 - Silicon Unilateral Switch (SUS): Really this is listed as a Diode, but is in actual fact a Thyristor.

25 - Static Induction Thyristor. (SITH): Very similar to a GTO: Gate Turn-off Thyristor.

26 - Thyristor Tetrode: A three electrode Silicon controlled switch with a Gate.

27 - TRIAC: Triode Alternating Current. The difference between this device and a DIAC which we talked about earlier is that the DIAC is an AC Unit which would be like two Zener Diodes back to back and would only let current flow if its breakdown voltage was reached. The TRIAC, in fact, has a switching Gate.

28 - Trigger DIAC: Used in conjunction with a TRIAC to provide a pulse to operate the Gate.

Note: I am sure that there may be more 'Types' of Thyristor hidden in the realms of electronics, but I have shown you a few different ones.

Misc. List of Equipment containing Diodes/Rectifiers/Thyristors:

Below are several units where their name does not fit into Diode, Rectifier or Thyristor, but in my opinion belong in this section: all in Alphabetical order.

1 - Composite Optical Isolator: See Optical Isolator.

2 - Composite Triode: Glass Tube Amplifier of the past about the size of a DVD player.

3 - Duo Triode: Vacuum Valves used in old radio sets.

4 - Ignitron: Gas filled Valves used as Controlled Rectifier

5 - Magnetic Optical Isolator: See Optical Isolator.

6 - Multi-Pixel Photon Counter (MPPC): See SiPM - Silicon Photo-Multiplier. Also known as a Silicon Photo-Multiplier. Can together with other equipment detect radiation.

7 - Optical (Opto) Coupler: See Optical Isolator.

8 - Optical Isolator: Comes in three types: Polarised, Composite and Magnetic. Other names are Optical (Opto) Diode in Diode section or Optical (Opto) Coupler. Generally used with galvanic systems and converts light to electronic signals of 4-20mA.

9 - Pentodes: These are Diodes in a way but they are television Vacuum Tubes of the past. Has five electrodes. Diodes have two electrodes, Triodes have three electrodes, Tetrodes have four electrodes and Pentodes have five Electrodes.

10 - Photo-coupler. Contains LED: Light Emitting Diode and a Photodetector. Identifies electrical signals and although uses internal light, does not emit light.

11 - Polarised Optical Isolator: See Optical Isolator.

12 - Reflex Klystron Oscillator: Used in microwave generators.

13 - Silicon Photomultiplier (SiPM): Sometimes called a MPPC: Multi-Pixel Photon Counter. See Multi-Pixel Photon Counter.

14 - Thyratron: This is a tube-shaped unit that is in fact a switch. Instead of a Vacuum Tube, in fact it is filled with Xenon or Hydrogen.

15 - Transferred Electron Device: See Gunn Diode Oscillator.

16 - Triode: An old fashioned Vacuum Tube type amplifier with three electrodes.

17 - Tetrode: An old fashioned Vacuum Tube with four electrodes. See Pentode.

Diode/Rectifier/Thyristor number of Electrodes:

| Diode
2 Electrodes | Triode
3 Electrodes | Tetrode
4 Electrodes | Pentode
5 Electrodes | Thyristor |

Sometimes the names of components are based on certain factors and the first four units above are based on the number of active electrodes that they have. Diodes and Rectifiers come in a whole range of designs, materials and colours besides the examples I have used above.

The Tetrode and the Pentode are, of course, older glass tube units with either a gas or vacuum inside. These glass tubes can have more electrodes as they may have heaters in them which require extra electrodes. Finally there is the Thyristor which, as stated, comes in many shapes and sizes and like the Triode may only have three electrodes.

Testing a Thyristor/SCR:

Testing Thyristors/SCRs in General:

Sometimes tests carried out on Thyristors/SCRs can seem to be ok, but unless approved methods are used then false readings may cause you to scrap perfectly good components.

The first thing to remember is to ISOLATE THE POWER SUPPLY before attempting to remove the Thyristor/SCR to be tested from the circuit mainly for your own safety, but also for other components which may be damaged.

How to test a Thyristor/SCR using a Multi-meter:

Remember in this instance you are using the Multi-meter Battery to provide the switching power. This is proven to be a reliable test method; there are others for instance with a battery and bulb.

Whatever test you perform, ensure that the Thyristor/SCR switching voltage is not exceeded which with a Multi-meter alone is highly unlikely.

Ensure that you take precautions if the voltage you are using is high enough to give an electric shock. Remember that short circuit arcs can damage eyesight!

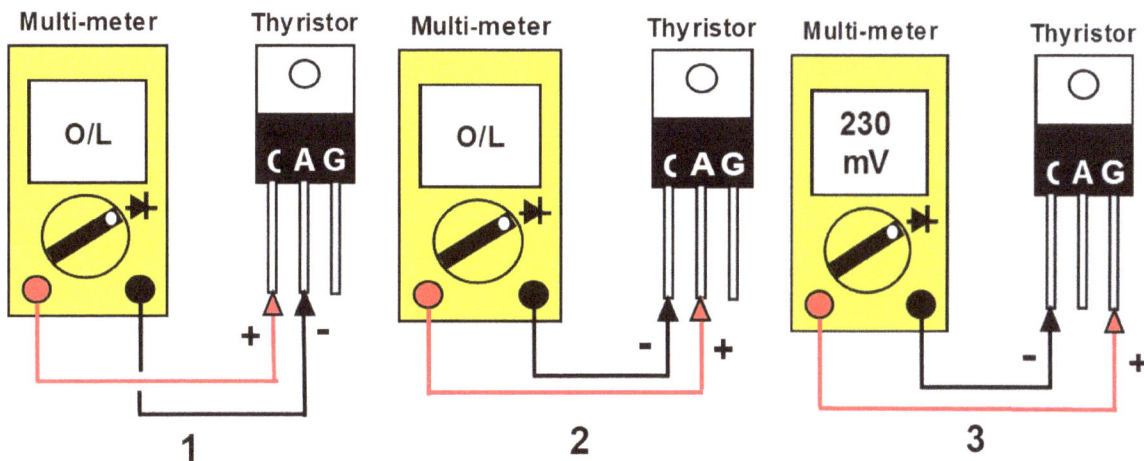

1) Select the Diode mode on the Multi-meter selector.

2) Connect the Positive Lead to Cathode (K) and Negative Lead to Anode as in Diagram 1.

3) Reading should be O/L.

4) Connect the Negative Lead to the Cathode (K) and Positive Lead to the Anode as in Diagram 2.

5) Reading should be O/L.

6) Connect the Negative Lead to the Cathode (K) and Positive Lead to the Gate as in Diagram 3.

7) Reading should be around 230mV.

8) Remove the Leads.

Now just a couple of tests to ensure that our Thyristor/SCR has returned to its rest state.

9) Connect the Positive Lead to Cathode (K) and Negative Lead to Anode as in Diagram 1.

10) Reading should be O/L.

11) Connect the Negative Lead to the Cathode (K) and Positive Lead to the Anode as in Diagram 2.

12) Reading should be O/L.

Diode Questions and Answers:

Q1 – What material goes to make a Diode?

A1 – MODERN DAY – Silicon or Germanium.

Q2 – Does a Diode have a volt drop?

A2 – YES. Silicon around 0.7 Volts older Germanium around 0.3 Volts.

Q3 – Why have Diodes changed to Silicon?

A3 – BREAKDOWN VOLTAGE – is one reason for the change to Silicon as these Diodes can handle much higher breakdown voltages.

Q4 – Would it matter if I connected a Standard Diode the wrong way round?

A4 – YES. A standard Diode would be in what is what is called reverse bias and just would not work, current would try to flow against the arrowhead symbol. If the current was high enough it would cause fatal damage inside of the Diode or it would explode! This could come under the term of PIV Peak Inverse Voltage.

Q5 – Are there Diodes that allow current to pass in reverse bias?

A5 – YES. Zener type Diodes allow current to pass when it reaches the 'breakdown' current of the Diode. These Diodes are used as an electronic safety valve.

Q6 – What does the arrowhead of a Diode represent?

A6 – CURRENT FLOW. The tip of the arrow is usually the cathode on a standard Diode.

Q7 – Why would a Diode get hot?

A7 – PURELY CURRENT.

Q8 – Does a Diode have a PN Junction?

A8 – MOST CERTAINLY. This is the reason they work.

Q9 – What is a Varactor Diode?

A9 – CAPACITOR. A Diode that performs as a Capacitor.

Q10 – Do standard Diodes have a volt drop?

A10 – YES depending upon the size of current at the junction.

Q11 – Is an LED classed as a Diode?

A11 – MOST CERTAINLY. LED stands for Light Emitting Diode. It uses a process where as the electrons pass into holes at the PN junction they emit light instead of heat, a process called electroluminescence.

Q12 – What is a Zero Bias PN Junction Diode?

A12 – NO VOLTAGE IS APPLIED – on the PN junction.

Q13 – Can I tell the Anode from the Cathode on a Diode?

A13 – YES. The Diode will have some method of distinguishing the Cathode. It may be a coloured band, a rim, or one electrode shorter than the other. In some cases, especially in large Diodes, a diagram of a Diode will be printed on the side.

Rectifier Questions and Answers:

Q1 – Is half wave Rectification ok for DC circuitry?

A1 – **NOT REALLY.** DC circuits usually require a full wave Rectification as half wave could be considered 'Noisy' in electrical terms.

Q2 – What is PIV Peak Inverse Voltage?

A2 – **MAXIMUM VOLTAGE** – in reverse bias that can be put onto a Diode or Rectifier without causing any damage.

Q3 – Sometimes a filter is used in a Rectifier circuit; why?

A3 – **PULSATING DC** – is sometimes the output result, the filter will smooth.

Q4 – Do modern Rectifiers involve Diodes?

A4 – **YES.** Most of them.

Q5 – Is a Bridge Rectifier full or half wave?

A5 – **FULL WAVE** – involves Diodes and inverts the negative part of the wave.

Q6 – Are there other ways to rectify full wave?

A6 – **YES** – with a Centre Tap Rectifier.

Q7 – What is a Precision Rectifier?

A7 – **A SUPER DIODE** – involving an Op-Amp (Operational **A**mplifier).

Q8 – Are Rectifiers in Selenium Stacks anymore?

A8 – **NO** – We have evolved into Silicon Semiconductors. Some collectors of older equipment may still have one, or a museum. The beauty of the stack Rectifier is that they could take high current. They tended in the past to be in large current battery chargers.

Q9 – Where did Vacuum Tube Rectification and Copper Oxide come in?

A9 – **OLDER RECTIFIERS** – like the Selenium we have evolved. Valves in old radios.

Q10 – What did a Vacuum Tube Rectifier look like?

A10 – **A VALVE IN AN OLD RADIO/TELEVISION.**

Q11 – What is the Ripple Factor?

A11 – **THE RATIO OF RMS AC INPUT TO DC OUTPUT.**

Q12 – Are there different types of Rectifier?

A12 – **YES.** I have listed thirty-one earlier and there may be more obscure ones.

Q13 – Why would a Rectifier get hot?

A13 – **CURRENT/FAULT.** Rectifiers can become overloaded if you give them too much work to do so they will get hot. They will also get hot if a fault is starting to manifest itself.

Thyristor Questions and Answers:

Q1 – What is a Thyristor?

A1 – A SWITCHED DIODE. I am sure that many electronic experts might debate this description with me. Like a Transistor, a Thyristor requires voltage onto a Gate to work.

Q2 – Is an SCR (Silicon Controlled Rectifier) the same as a Thyristor?

A2 – YES. Also referred to as a Thyroid Transistor.

Q3 – Is a Thyristor a 'stable' device?

A3 – YES. Both open and closed.

Q4 – What is Thyristor Switch on Voltage?

A4 – THE MINIMUM VOLTAGE – that the Thyristor can operate.

Q5 – So what is the Minimum Control Electrode Current?

A5 – THE MINIMUM CURRENT – that is required to switch the Thyristor on.

Q6 – What is 'Maximum Allowed Forward Current?

A6 – MAXIMUM current the Thyristor can take.

Q7 – Are the electrodes marked as to 'Gate' etc.

A7 – YES. Usually on the side of the Thyristor.

Q8 – Did smaller Thyristors used to be round and flattish?

A8 – YES – but components have evolved. Depends upon the date it was manufactured.

Q9 – I noticed some Thyristors are square plastic, but why is metal protruding with hole?

A9 – HEAT SINK – Thyristors, by nature of how they work, will produce heat.

Q10 – Can I test the Thyristor with a multi-meter.

A10 – YES – but not as easily as other components. You will have to make up a test circuit as shown earlier.

Q11 – If the voltage is removed from the Gate does the Thyristor close immediately?

A11 – NO – not until the Anode voltage falls to below the 'Cut Off' Voltage!

Q12 – Can Thyristors be obtained that latch?

A12 – YES. The majority do and this is where they differ from a Transistor.

Q13 – How long does it take for a particular Thyristor to open and close or switch.

A13 – MANUFACTURERS can supply this information which might be critical when designing the circuit.

Q14 – Has a Thyristor always got three electrodes?

A14 – YES. Anode, Cathode and Gate, similar to a Transistor.

Capacitor Section:

A Capacitor is a passive device that stores energy; passive because it uses energy but does not produce energy. Different from a battery which uses a chemical reaction to produce energy as well as store it.

This component is just two plates, one with a positive charge and one with a negative charge separated by a dielectric. They can be 'Fixed' Capacitors, which are probably the most common, or 'Variable' Capacitors.

As far as Capacitors go, the Ceramic Capacitor is one of the most common on a PCB, these are round, resembling a button, with two electrodes. These, along with several others, are explained later in this section.

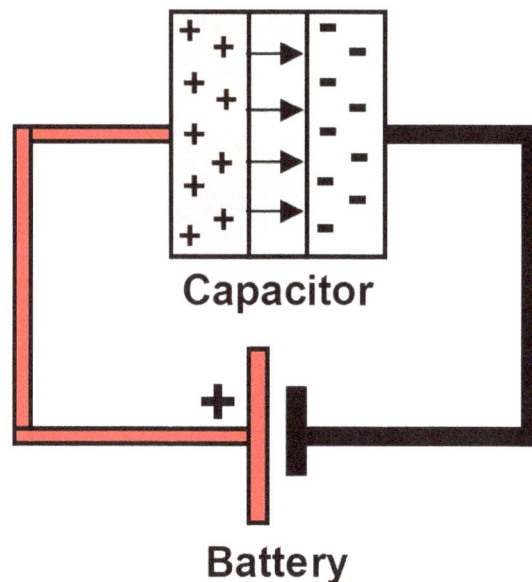

There are two main types of Capacitor:

Polarised: These are polarity conscious and must never be connected to a power supply in reverse bias i.e. Positive of Battery onto Capacitor Negative. Doing so can cause catastrophic damage to the Capacitor or cause it to explode. Usually the Positive (+) and the Negative (-) are marked on the side.

Non Polarised: These can be connected any way round.

Capacitors are marked in Farads and Capacitor testing is discussed later.

As we go through this section we will have a look at the different capacitors, how they work, what they are used on and how to test them.

NOTE:

ON ALL ATEX EQUIPMENT FOR HAZARDOUS AREAS SUCH AS ZONES, REMOVAL/CHANGING OF CAPACITORS FROM A MANUFACTURERS' PRINTED CIRCUIT BOARD CAN ONLY BE CARRIED OUT WITH MANUFACTURER APPROVAL IN WRITING.

FAILURE TO COMPLY WITH THIS MAY BE CLASSED AS AN UNAUTHORISED MODIFICATION AND MAY RENDER ANY CERTIFICATION DOCUMENTATION NULL AND VOID.

ALSO THE DESIGN, MANUFACTURER, TYPE ETC., OF THE CAPACITOR COMPONENTS MUST NOT BE CHANGED IN ANY WAY.

Capacitors:

How do Capacitors work?

So what is the purpose of a Capacitor? Firstly, Capacitors are passive components, which means that they use energy and do not produce it. Capacitors store charge but are unlike a battery which is not a passive component, as although it stores charge, it also produces that charge chemically. Capacitors can be 'Through Hole' with protruding long electrodes, or 'Surface Mount'.

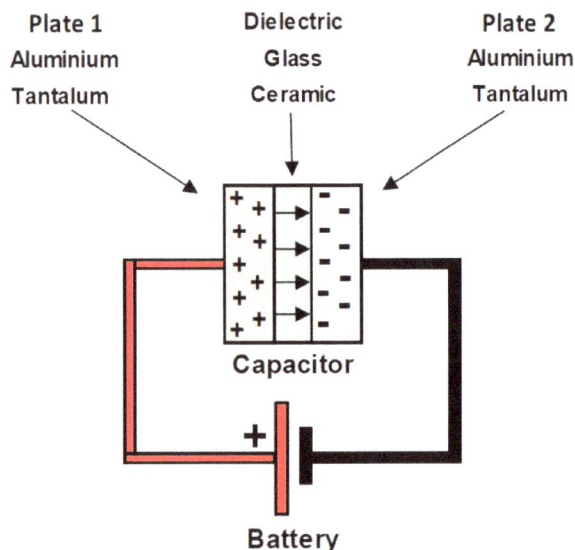

Plate 1	Dielectric	Plate 2
Aluminium	Glass	Aluminium
Tantalum	Ceramic	Tantalum

Capacitor

Battery

They store electric charge on plates that are insulated away from each other by a material called a 'Dielectric' as per the diagram on the left. I have given a couple of examples of plate metals and dielectric.

The battery causes electrons to flow onto the plates, but the dielectric restricts that movement from one plate to another so the charge remains on the plates.

The charging from the battery will produce a high current at the start and will slowly drop off. The charging will completely stop when the voltage that has built up on the plates equals the battery voltage.

The Capacitor will hold its charge for a while, but unless discharged the charge will slowly leak away.

Capacitor Kinds & Types:

Firstly let me say that basically there are two, of what I call, 'KINDS' of Capacitor, namely:

1) **Fixed:** Where the Capacitor plates are a fixed distance apart so it will store a fixed amount of charge and these two parameters do not vary. So the value may be displayed on the body; or

2) **Variable:** Where the plate distance apart can be adjusted and so will adjust the capacitance and charge that can be accepted.

I will go on to discuss several 'TYPES' of Capacitor that come under these two headings later in this section.

Capacitor Sizes:

The size of a Capacitor is measured in Farads; common examples are below.

1)	Kilofarad	= kF	(one thousand)	1000F
2)	1 x Farad	= F	(one)	1F
3)	Millifarad	= mF	(one thousandth)	0.001F
4)	Microfarad	= μF	(one millionth)	0.00,0001F
5)	Nanofarads	= nF	(one billionth)	0.000,000,001F
6)	Picofarad	= pF	(one trillionth)	0.000,000,000,001F
7)	Femofarads	= fF	(10^{-15})	0.000,000,000,000,001F

Fixed Capacitors:

Firstly some Capacitors can be further separated into two further, what I have called, **'GROUPS'** namely:

1) **Polarised,** these have two terminals and are polarity conscious and must be connected correctly as they only work in one direction and can explode if connected incorrectly!

2) **Non-Polarised,** where they can be connected either way round, possibly the most common. **Connecting a Polarised Capacitor the incorrect way round can be very hazardous!**

Fixed Capacitors include types like **1)** Ceramic Capacitors **2)** Film Capacitors **3)** Electrolytic Capacitors **4)** Air Capacitors **5)** Pseudo Capacitor etc.

1) Ceramic Capacitors (Fixed Value, Non-Polarised):

On printed circuits, at first glance, these can get mixed up in looks with a Non-Linear Resistor. They can come in all shapes and sizes. These Capacitors are **Non-Polarised** so are not polarity conscious and can be connected either way round. They can be Leaded Disc or Multi-layer.

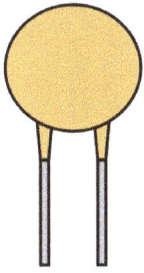

So what are these Capacitors made of and how do they do their job? Well, looking at the diagram to the left, these are 'Disc' Capacitors. The top orange disc of the Capacitor in this case would contain Metal Plates and Ceramic Dielectric Material.

Ceramic Capacitors are two plates separated by a Dielectric, a Dielectric being an insulating type material that is a poor conductor of electricity with no free electrons. They can be multi-layered but these are usually used in higher voltage situations.

There are two types of electrode configuration. There is the **'Radial'** electrode as in the above diagram where both electrodes of a two electrode Capacitor come out of the bottom, or the **Axial** electrode configuration where the electrode comes out of either end of a cylinder Capacitor.

Materials for the Dielectric include: Silicon, Mica, Titanium Dioxide, Barium Titanate etc. These come in four 'Classes' depending on what level you want the dielectric to be at.

As stated above, the Ceramic material itself can be something like Silicon, Porcelain etc. These Capacitors can go up to quite high voltages, in the kV region, and as above can be a single Layer or Multi-layered (MLCCs). Layers can be in the hundreds!

Ceramic Capacitance Value:

Looking at Resistors you will see that the value of the Resistor in Ohms (Ω) is achieved by coloured bars on the Resistor or done by writing on the side of larger ones. Ceramic Capacitor value is achieved as below:

They can range from µF to pF the value usually being printed onto the face of the disc. If the value is three digits then the first two digits are the Capacitance value and the third number is the multiplier (number of 0s); remember these Capacitors are usually in Pico-farads pF. Looking at the diagram to the right there is the number 204, so this is telling us that the capacitor is 20 and 4 zeros = **200,000pF** or **200nF.** The Letter is the tolerance (Below) **+/- 5%.**

Ceramic Capacitance Tolerance:

Looking at Resistors, the Tolerances of the values are worked out by coloured bands, but with Ceramic Capacitors that is slightly different. Below there is a chart which will correspond with a particular letter which is stamped on the face of the Ceramic Capacitor.

A	± 0.05pF		K	± 10%
B	± 0.1pF		L	± 15%
C	± 0.25pF		M	± 20%
D	± 0.5pf		N	± 30%

E	± 0.50%		P	- 0% +100%
F	± 1%		S	- 20% +50%
G	± 2%		W	- 0% +200%
H	± 3%		X	- 20% +40%
J	± 5%		Z	- 20% +80%

The tolerance spreadsheets for the UK are shown in the diagrams to the left, but with European markings the values are completely different although the lettering looks similar.

Even the Capacitor value markings are different on the European system if you ever have anything to do with it.

Film Capacitors are different again.

I hope that this has not confused the issue, but one day you may require the information.

The drawing symbols for this type of Capacitor are shown in the diagram to the right.

International Capacitor Colour Coding System:

International Capacitor Colour Coding									Example:
Colour:	Code A:	Code B:	Multiplier:		Colour:	Code A:	Code B:	Multiplier:	
Black	0	0	x1		Green	5	5	100,000	
Brown	1	1	x10		Blue	6	6	1,000,000	
Red	2	2	x100		Violet	7	7		
Orange	3	3	x1000		Grey	8	8	x0.01	
Yellow	4	4	x10,000		White	9	9	x0.1	
Silver			x0.01		Gold			x0.1	26,000pF or 26nF

This system is an older one and modern systems use numbers and letters but you may come across an older Capacitor with a colour code. Voltage & Tolerance may be displayed as extra bars.

2) Film Capacitors (Fixed Value, Non-Polarised):

This Capacitor, which is probably the most common in printed circuits, gets its name from the thin 'Film' of Polyester, Polypropylene or Polyethylene Terephthalate Dielectrics used in its construction. Because of the 'Poly' type Dielectric these are sometimes known as 'Plastic' Capacitors.

These Capacitors are Non-Polarised meaning that they are not polarity conscious and can be connected any way round in a circuit.

They are said to be 'self-healing', meaning that in the event of an overload they may still work because of the make-up of the electrodes and interior.

There are two types of Film Capacitor, namely:

1) Film Foil and 2) Metallised Film.

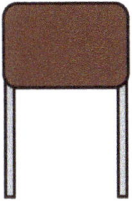

Film Foil Capacitors use very thin sheets of Plastic Dielectric with sheets of Aluminium sandwiched between them which are the Electrodes. These Capacitors can cope with fairly high currents.

Metallised Film Capacitors are slightly different, here the Plastic Dielectric has a metallic coating sprayed actually on its surface which are the electrodes.

Another metal that can be used for the electrodes is Zinc although the Aluminium metal is more common. The Dielectric may be thicker in the Film Foil Capacitors and a bit thinner in the Metallised Film Capacitors.

Types of Film Capacitor:

KT	**Foil/Polyester**
MKT	**Metallised Polyester**
KP	**Foil/Polypropolene**
KC	**Foil/Polycarbonate**
MKC	**Metallised Polycarbonate**
KS	**Foil/Polystyrene**
KPS	**Foil/Polyphenylene Sulphide**
MPS	**Metallised Polyphenylene**
MPS	**Metallised Paper**

The letters on the spreadsheet to the left may determine on older Film Capacitors what the internal materials actually consist of, which would give a good idea as to what type of Capacitor you are actually dealing with.

Again, voltage plays a huge part when selecting the type of Capacitor in your circuit.

The question arises in that if you are designing a circuit which type of Capacitor do you use? Consultation with manufacturers is always a good idea. You require Farad Value, Tolerance and Voltage for a start.

The drawing symbols for this type of Capacitor are shown in the diagram to the right.

3) Electrolytic Capacitor (Fixed Value Polarised):

These Electrolytic Capacitors come in all shapes and values and can be obtained Axial. Here instead of a straight Dielectric this Capacitor uses a similar material called an Electrolyte in the form of a Polymer or a Gel.

The Electrolyte contains many more 'Ions' than the Dielectric normally used in many Capacitors so can form a higher capacitance. Electrolytic Capacitors are 'Polarised' which means that they have an 'Anode' and a 'Cathode' and must be connected the correct way round in a circuit.

Positive and Negative may be marked on the body of the Capacitor. The most common models are as per the diagram below:

Being 'Polarised' means that the voltage of the circuit must always be greater on the Cathode than the voltage on the Anode. The most common metals used here are Tantalum or Aluminium. One problem with an Electrolytic Capacitor is that we can get Capacitor 'Drift' over time and can drift as much as 20% away from their nominal value.

A big drawback is that, being polarised, if this Capacitor is connected the wrong way round or for any reason the voltage on the Anode becomes higher than the Cathode then this could cause the Capacitor to overheat and at the worst, explode!

When this Capacitor is being manufactured the Aluminium Electrodes, the Anode being coated with an oxide, are sandwiched in with the Electrolyte and rolled into what may resemble a 'Swiss Roll', which gives it the shape shown in the diagram above.

Usually the Capacitor value will be marked onto the body, usually in μF (Micro-farads) along with the voltage. It is imperative that the voltage on the side of the Capacitor is never exceeded otherwise this can also cause the Capacitor to explode.

If the value is three digits then the first two digits are the Capacitance value and the third number is the multiplier (number of 0s).

If the value of the Capacitor, say in μF, is preceded by a letter as well as the three numbers, then this is the voltage. Some Capacitors use this method instead of actually printing the voltage onto the body.

i.e. A=10Volts, C=16Volts, D=20Volts, E=25Volts, G=4Volts, H=50Volts, J= 6.3Volts, V=35Volts etc.

The drawing symbols for this type of Capacitor are shown in the diagram to the right. Being polarised has a plus sign.

4) Hybrid Capacitors:

Hybrid Capacitors come in all sorts of designs, the diagram below is just one type. They are a form of what we call today 'Super' Capacitors. They combine all of the best qualities of many Capacitors into one.

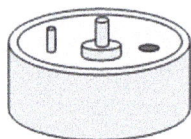

Sometimes these advanced Capacitors use a liquid electrolyte combined with a polymer. They have a distinct stable & reliable capacitance. On a standard Capacitor, sometimes its temperature coefficient is not so good and the capacitance can 'drift' but not in this case.

5) Pseudo Capacitor

When I said at the beginning of this section that Capacitors and Batteries both store charge but in a different way, well a Pseudo Capacitor can be called a cross between a Capacitor and a Battery.

This is a type of Super Capacitor sometimes called a Faradaic Capacitor. It has an electrolyte based on Redox (**Red**uction **Ox**idation) which is to do with Atomic electron flow between the plates and through the electrolyte.

These Capacitor chemicals include Metal Sulphides, Hydroxides, Oxides & Nitrides along with Polymers.

Several types of Fixed Capacitor:

We have looked at several common Capacitors and how they operate, but there are many more Capacitors, some of which we will see overleaf and there may even be more.

Sometimes the Capacitor will have the same name i.e. Electrolyte/Dielectric and will be a different shape from the ones below. It sometimes all depends upon the manufacturer of the duty of the Capacitor and whether it is cylindrical or parallel plate.

Ceramic Disc Capacitor	Surface Mounted Ceramic Capacitor	High Voltage Ceramic Capacitor	Axial Feedthrough Ceramic Capacitor
Polycarbonate Capacitor	Polyester Capacitor	Polystyrene Capacitor	Polypropylene Capacitor
Motor Run Capacitor	Light Emitting Capacitor	Glass Capacitor	Paper Capacitor
Silver Mica Capacitor	Vacuum Capacitor	Silicon Capacitor	Niobium Capacitor
Tantalum Pearl Capacitor	Tantalum Chip Capacitor	Aluminium Electrolyte Capacitor	Super Capacitor
Tantalum Chip Capacitor	Monolithic Capacitor	Suppression Capacitor	Wire Ended Capacitor

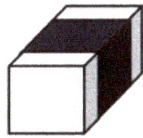

Three Terminal Capacitors:

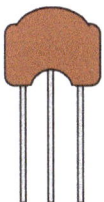

Before we go onto Variable Capacitors I would just like to mention the Three Terminal Capacitor. This Capacitor could be, for instance, a Multi-Layer Ceramic Chip Capacitor; we have talked about Ceramic types earlier in this section. The third electrode of this Capacitor is connected to earth.

These Capacitors can be 'Through Hole' like the diagram on the left, or a Surface Mounted Capacitor.

Variable Capacitors:

The main way of changing a capacitance is to vary the distance between the plates and that is exactly what a Variable Capacitor does. As in other Capacitors there are two plates, one is fixed and one is movable with the adjuster. As in an electric motor the fixed (non-movable) plate is called the 'Stator' and the movable plate is called the 'Rotor'.

There are two types of dielectric that are used with Variable Capacitors, namely air or solid. I will just mention that it is possible to obtain 'Polarised' Variable Capacitors.

1) Standard Variable Capacitor:

Sometimes in certain circuits it would be ideal if the resident Capacitor could be adjusted in some way so as to give the exact capacitance required. Variable Capacitors can be, as the name suggests, varied within certain parameters. Sometimes used in situations where tuning is involved in, say, a radio receiver.

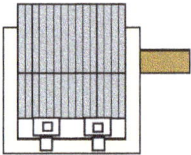

So how do we change the capacitance? Well what we would need to do is vary the distance between the plates and that is exactly what happens here.

The plates are adjusted by a hand rotation shaft, which alters the plate position. There is a 'fixed' plate that never moves and the adjustable moving plate called the 'rotor' plate.

There are two types of dielectric that can be used with these Variable Capacitors and they are 1) Air, which can be single or double connected, and 2) Solid, which can be, as we have mentioned in the Fixed Capacitors, a Poly plastic material or Mica.

2) Trimmer Capacitor (Variable Capacitor):

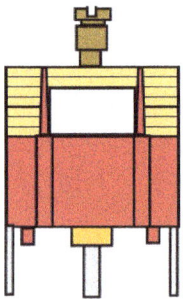

Although these trimmer Capacitors are a type of Variable Capacitor they are used more for Calibration than a designated circuit Capacitor.

So here we have the adjusting screw on the top screwing into the 'Steatite' base of the Capacitor inside and again moving the plates together and apart to alter the capacitance. A Silver fixed plate being the 'stator' plate and the moving plate being the 'rotor' plate – very similar to a Variable Capacitor.

The solid dielectric which is between the plates could be Titanate Ceramic.

3). Tuning Capacitor (Variable Capacitor):

This Tuning Capacitor is again a type of Variable Capacitor very similar to the Trimmer Capacitor in its operation. Used in Radios,

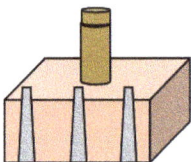

This Capacitor internal has two separate sides, one for AM Radio and one for FM Radio. Each side consists of two variable Capacitors. These Capacitors are not so common these days and there are many designs; my diagram to the left is one design.

The drawing symbol for this type of Capacitor is shown in the diagram to the right. Sometimes the Symbol is turned 90 degrees.

Dielectric Materials:

So as we have mentioned, the dielectric is the insulator material between the Capacitor plates. When we ask exactly what causes one dielectric to be better than another we must look at its permittivity. The better the permittivity the more effective the Capacitor.

Although an insulator, the dielectric surface close to the plates does become polarised with a positive and negative charge because of the field set up by the plates.

The dielectric needs to be completely in contact with the Capacitor plates, any gaps will fill with air and could reduce the capacitive value of the Capacitor.

I have put air as a dielectric below and this is very close to a baseline of 1, but by manufacturers using other materials with a better permittivity **(the ability to store electrical charge)** than air, which is close to 1, the capacitance can be increased enormously! So the permittivity of the dielectric will be a number greater than 1.

Below is a list of several dielectric materials that are used, some I am sure that you will have heard of and will be fairly common; others may be a bit more obscure and I am sure there will be many more:

1) Glass	13) Bakelite	25) Steatite
2) Vacuum	14) Paper	26) Cellulose
3) Polyimide	15) Mica	27) Fibre
4) Polyethylene	16) Quartz	28) Plexiglass
5) Polypropylene	17) Porcelain	29) Mycalex
6) Polystyrene	18) Tantalum Pentoxide	30) Alsimag
7) Polyester	19) Ceramic (Silicon Nitride)	31) Mineral Oil
8) Poly Vinyl Chloride	20) Pyrex	32) Rubber
9) Dry Air	21) Plastic Film	33) PTFE/Teflon
10) Oak Wood	22) Barium Titanate	34) Silicone
11) Birch Wood	23) Strontium Titanate	35) Nylon
12) Maple Wood	24) Titanium Dioxide	36) Shellac

Capacitor Plate Materials:

As mentioned earlier, Capacitors can be all sorts of shapes and sizes, but the plates mainly boil down to two, namely: 1) Cylindrical; or 2) Parallel Plate. Parallel Plate speaks for itself, i.e. two plates in parallel, but with the Cylindrical type they firstly trap the metallic film either side of the dielectric in a long sandwich, they then roll the sandwich very much like a stick of rock into a cylinder.

The charge builds up on the plates when an electrical supply is put onto them. Making the plates larger or moving the plates closer together will increase the capacitance.

Below is a list of several plate materials that are used, some I am sure that you will have heard of and will be fairly common, others may be a bit more obscure and I am sure there will be many more:

1) Silver Plate	5) Tantalum	9) Zinc Alloy
2) Silver Coating	6) Titanium	10) Tin Foil
3) Etched Aluminium Foil	7) Niobium	
4) Formed Aluminium Foil	8) Copper	

Capacitor Calculations:

Sometimes even on printed circuit boards Capacitors are connected in series or parallel. So if they are connected in either of these combinations how do we arrive at the total? Capacitors may be placed in parallel to provide a higher level of Capacitance.

Capacitors in Parallel:

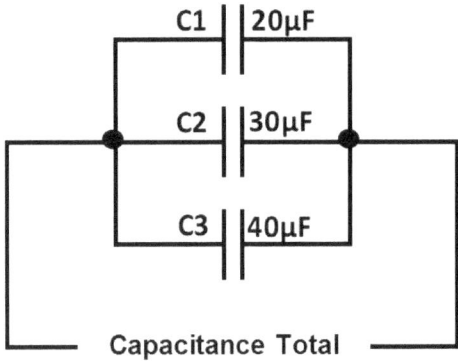

To calculate Capacitors in Parallel it is the total opposite to Resistance:

Here in parallel we simply add up all of the individual Capacitor values together:

Capacitance Total	=	C1 + C2 + C3
Capacitance Total	=	20µF + 30µF + 40µF
Capacitance Total	=	90µF

Two Capacitors in Series:

To calculate two Capacitors in Series it is the total opposite to Resistance:

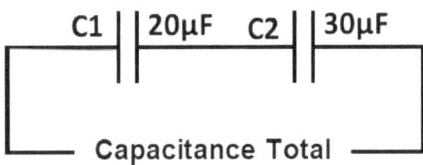

For 2 Capacitors in Series we use the formula:

$$\text{Capacitance Total} = \frac{C1 \times C2}{C1 + C2}$$

$$\text{Capacitance Total} = \frac{20µF \times 30µF}{20µF + 30µF}$$

$$\text{Capacitance Total} = \frac{600µF}{50µF} = 12µF$$

Three Capacitors in Series:

To calculate three Capacitors in Series it is the total opposite to Resistance:

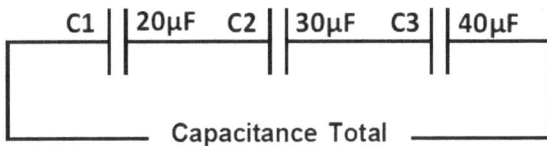

For 3 Capacitors in Series we use the formula:

$$\frac{1}{\text{Total}} = \frac{1}{C1} + \frac{1}{C2} + \frac{1}{C3}$$

$$\frac{1}{\text{Total}} = 0.05µF + 0.033µF + 0.025µF$$

$$\frac{1}{\text{Total}} = \frac{1}{20µF} + \frac{1}{30µF} + \frac{1}{40µF}$$

$$\frac{1}{\text{Total}} = \frac{1}{0.108} = 9.26µF$$

The Parallel and Series formula works regardless of the number of Capacitors in the circuit. Just remember that the calculations for Capacitors are the direct opposite of those for Resistors.

Testing a Capacitor:

When we talk about testing a Capacitor, what actual tests to we have to do and what instrument do we use for the testing? We have talked about Capacitors charging up to full capacity so that the voltage on the plates equals the charge voltage. Just be careful that the instrument voltage does not charge up the capacitor.

That Capacitor charge voltage may be used, for instance, in a bridge rectifier circuit to cause the inverted sine wave peaks to be closer to a straight line. Another use may be in a reacceleration system to assist in slugging the contactor.

Firstly we must ensure for safety that the Capacitor is fully discharged. I have seen technicians put their screwdriver across the Capacitor terminals. This is not advisable as there will be a loud crack and a spark. The resultant surge inside the Capacitor could damage its plates.

A Discharge Resistor connected like the one in the diagram to the right is always the best way as this can be done without damage.

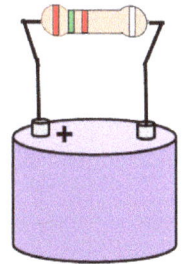

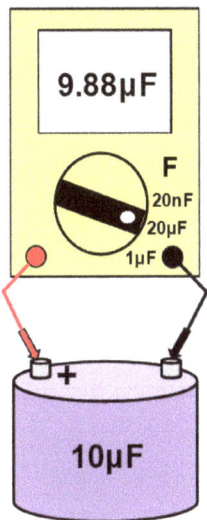

So we have decided to test the Capacitor using a multi-meter with Capacitor Test Facility. On the left there is a diagram of a multi-meter with a Capacitor Test Facility. Turn the adjuster to 'F' (Farads) for capacitance and this will have various scales, sometimes more than the amount as shown.

The meter may also have a section where you can plug the electrodes of the Capacitor into the meter without using the red and black leads. Before testing can begin, as noted above, you must ensure that the Capacitor is discharged.

As shown in the diagram the reading of 9.88μF on the meter is very close to the value on the side of the Capacitor of 10μF and this is a good sign that the Capacitor is ok. If there was a large discrepancy between what the meter was reading and the value on the side of the Capacitor then it may be faulty. The reading on the meter may be more accurate than the one on the Capacitor.

If the Capacitor, as in the picture, is polarity conscious (Polarised) and the leads are used on the multi-meter then the red must go to the Capacitor positive.

Let us suppose that your multi-meter has not got the Capacitor testing facility, how can you tell if a capacitor is ok? Well we can use the resistance setting.

Put the multi-meter on Ω – MΩ, using the meter test leads as in the diagram to the right, put them across the Capacitor.

One reading we are **NOT** looking for is 0Ω as this would indicate that the Capacitor has gone **'short'** circuit and is defective.

Another reading we are **NOT** looking for is something like **5MΩ** as this would indicate that the Capacitor has an **'open'** circuit.

Ideally a reading of several thousand Ohms would probably indicate that the Capacitor is healthy.

Testing Capacitors with resistance is a bit hit and miss and may involve a bit of guesswork as to the readings.

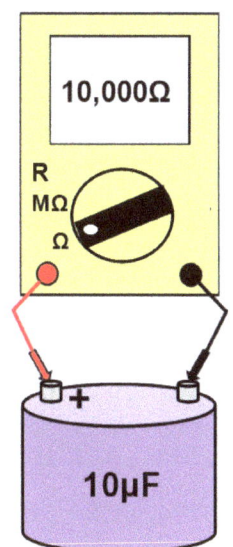

Measuring the Capacitance:

The capacitance of a Capacitor can be measured by the formula: $Q = CV$

Q (Total) = C (Capacitance) x V (Voltage) so we can say that the voltage applied will have a direct influence on the total. This of course is caused by the voltage influencing the charge.

Capacitor Questions and Answers:

Q1 – How do Capacitors store energy?

A1 – IN AN ELECTROSTATIC FIELD.

Q2 – How do Inductors store energy?

A2 – IN A MAGNETIC FIELD.

Q3 – How is the Capacitor voltage rating determined?

A3 – THE DIELECTRIC INSULATION STRENGTH.

Q4 – What is Capacitive Reactance?

A4 – THE CAPACITORS OPPOSITION TO AC.

Q5 – What is Capacitance Measured in?

A5 – FARADS.

Q6 – Is an Electrolytic Capacitor a Polarised Capacitor?

A6 – YES.

Q7 – Is a vacuum classed as a dielectric?

A7 – YES, and a very good one.

Q8 – Are Capacitors colour coded as to their value?

A8 – SOME OF THEM – others have value in numbers & letters on them.

Q9 – Are Capacitors colour coded for voltage?

A9 – AGAIN, SOME OF THEM.

Q10 – Can Capacitors be tested with a Multi-meter?

A10 – YES – some meters have a Farad Test Point, but they can be tested in Ohms (Ω).

Q11 – Can Capacitors fail short circuit?

A11 – THEY CERTAINLY CAN.

Q12 – Are there different dielectrics?

A12 – YES – I have earlier listed thirty-six. There may be even more.

Q13 – Are there different Capacitor Plate materials?

A13 – YES – I have listed ten. There may be more.

Q14 – Can Capacitors be variable?

A14 – YES – I have listed three. There may be more.

Q15 – Can a Capacitor have three electrodes?

A15 – YES – Look at the earlier sections.

Q16 – Is a Capacitor a passive component?

A16 – YES – it uses energy and does not produce it.

Resistor Section:

Resistors are one of, if not **'the'** most common components on a printed circuit board and there are many types, with possibly Carbon Film being the most common. They can be linear, non-linear, variable etc. Mainly used for things like Current Control.

Resistors are passive components, which means that they use energy and do not produce it. This section talks about how to distinguish between the different types of Resistor and how to determine its value.

Some Resistors have coloured bands on them and we discuss easy ways to remember the colour value and also the tolerance value. Test methods are discussed on using a Multi-meter to determine and check the value of the resistance.

There is a section on how to calculate the resistance of several Resistors connected in parallel and series.

The Resistor above is a common Linear Resistor so let us discuss in this section how to determine its values and those of many other Resistors by looking at the information on the case and also discuss the test methods available.

In an analogy: just imagine a situation where the electrons are in a race; moving over solid ground is easy, but introduce some resistance and it would be like making them run through inches of water which would be not quite so easy. Introduce a lot of resistance and it would be like making them run through treacle!

Resistors come in many forms as you will see as we go through this section. The most common fitting for printed circuits is probably the 'through hole' (PTH – Plated Through Hole) where the Resistor has two long electrodes. They can be surface mounted which would usually be an oblong shape with the resistance marked on.

Remember that the common Carbon Film Resistor is low wattage. We are looking at 0.25 Watts up to 3 Watts. Any higher and a different type of higher wattage Resistor must be used.

NOTE:

ON ALL ATEX EQUIPMENT FOR HAZARDOUS AREAS SUCH AS ZONES, **REMOVAL/CHANGING OF RESISTORS FROM A MANUFACTURERS' PRINTED CIRCUIT BOARD CAN ONLY BE CARRIED OUT WITH** MANUFACTURER APPROVAL IN WRITING.

FAILURE TO COMPLY WITH THIS MAY BE CLASSED AS AN UNAUTHORISED MODIFICATION **AND MAY RENDER ANY CERTIFICATION DOCUMENTATION NULL AND VOID.**

ALSO THE DESIGN, MANUFACTURER, TYPE ETC., OF THE RESISTOR COMPONENTS MUST NOT BE CHANGED IN ANY WAY.

PCB Non Linear Resistors:

Resistor Kinds & Types:

So what is the purpose of a Resistor? Firstly, Resistors are passive components, which means that they use energy and do not produce it. They are there to regulate or stem the flow of Electrons by the mere make-up of what the Resistor is made of.

Firstly let me say that basically there are two **'KINDS'** of Resistor namely: **1) Linear** and **2) Non-Linear.** I will go on to discuss several **'TYPES'** of Resistor that come under these two headings.

Non-Linear Resistors:

Examples of **Non-linear** Resistor types are components such as **1)** Thermistors, **2)** Varistors, **3)** Photo Resistors and **4)** Surface Mounted Resistors and **5)** Memristor. Let us look at these five Resistors as follows:

1) The Thermistor (Non-Linear Resistor):

I am sure that you have heard of the Thermistor before today, but where would you find one in industry and how do they work?

A Thermistor may look like the diagram to the left with a body and two electrodes. This is a very temperature sensitive semiconductor device, its resistance relies on temperature. The actual name Thermistor comes from the words Thermally Sensitive Resistor.

This device is further divided into two types: 1) Positive Temperature Coefficient (PTC) where its resistance **INCREASES** with an **INCREASE** in temperature and 2) Negative Temperature Coefficient (NTC) where the resistance **DECREASES** when the temperature **INCREASES.** The change in resistance will follow a non-linear curve which can be plotted.

So you must decide whether you want a PTC or NTC component, and when you purchase it from the manufacturers, the Thermistor will start off with a certain nominal value of resistance and they will probably provide you with a diagram of the non-linear curve.

The main duty that my company used Thermistors for was certain electric motor winding temperatures. Thermistors were put into the windings of the motor, by the motor manufacturers, and a cable, separate from the motor 'load' cable, would go from the motor and link into the motor control unit i.e. the starter.

If the winding temperature started to rise beyond certain parameters then this component would sense the rise and trip the motor. Other uses are refrigerators & ovens.

So what are these Thermistors (**Non-Linear Resistors**) made of? Usually they are made of Metallic Oxide semiconductor material encapsulated in an outer shell made of another material such as a resin. The Metal Oxide, together with other added interior chemicals, forms a PN Junction between them.

2) Varistors (Non-Linear Resistor):

I am sure that you may have heard of a Varistor before now as these are not an uncommon component; they are sometimes called a Voltage Dependant Resistor. (VDR)

Apart from the similarity of looking like a different coloured Thermistor, the Varistor has two electrodes and works as an over voltage/surge Semiconductor device and not temperature. This component is looking for peaks in voltage which dramatically changes its resistance.

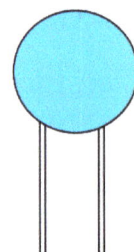

So under normal operation this Resistor would be a HIGH resistance. If the voltage was to go outside of its parameters then its resistance would decrease very rapidly caused by an **'Avalanche'** effect in the atoms of the Semiconductor as discussed in the Diode section. The decrease in resistance would absorb the overvoltage.

So what would be its position in the circuit? Well, it would be connected in parallel with the electrical circuit that it is protecting, the object being as above to absorb any over voltage that may damage the circuit. Zener Diodes, in the Diode Section, achieve a very similar objective in a slightly different way.

The most common Varistor is again a Metal Oxide interior, but Metrosil manufactured a Varistor back in the 1970s which we used on our substation instrumentation, that was a Silicon Carbide Varistor.

The Metal Oxide, together with other added interior chemicals, forms a PN Junction. The current flow curve is obviously **Non-Linear**. These Resistors seem to defy Ohm's Law if you think about it.

3) Photo Resistor (LDR) (Non Linear Resistor):

Many Technicians will remember this type of Non-Linear Resistor from the photocell. Another name is the **'Light Dependant Resistor'.**

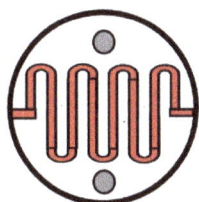

You may recognise the design of the Resistor in the diagram to the left. Its resistance changes dramatically when exposed to light and dark! Where does the name come from? Well, besides 'Photocell' the name comes from 'Photon' and 'Resistor'.

There are two types of light Dependant Resistors namely 1) Intrinsic and 2) Extrinsic, the difference being their Semiconductor content.

Firstly let us look at 'Intrinsic' Resistors; these are usually made from pure, un-doped (nothing added) Semiconductor materials such as Silicon or Germanium and in this case, when Photons (Sunlight) lands onto the element, they excite the Atoms in the Electrolyte and cause electrons in their outer shell or valence to jump orbits.

Extrinsic Resistors, however, have their Electrolyte doped with other materials causing a much easier Atom excitation and cause electrons to jump orbits from the shells above the valence.

The resistance will change depending upon how much light falls on the photocell, hence in the light it will be low resistance and in the dark, high resistance. So we can now say that this Resistor is Non-Linear.

4) Surface Mounted Resistor (SMD) (Non-Linear Resistor):

These could be termed as **'Integrated' Resistors** as the Resistor is integrated onto a Ceramic Chip and are sometimes also referred to as **'Chip' Resistors**. These are preferred as the Resistors on printed circuits and are very distinctive as per the diagram below.

So looking at the diagram to the right we have a black ceramic body with electrodes on either end. Next, a resistive material is literally printed onto the ceramic surface. The Resistor is then covered in various protective coatings to protect it against any damage.

So how do we tell what resistance value these Resistors are? Well usually on each component is a three figure number, the first two digits are the resistance value and the last figure is the number of zeros. So a number 452 would be 4500Ω but you might require a magnifying glass to read it.

5) Memristor (Non-Linear):

This is a semiconductor device, the name 'Memristor' coming from the name **Memory Resistor**. This Resistor remembers the resistance it had when a current last flowed through it, so it can be programmed to whatever value is required.

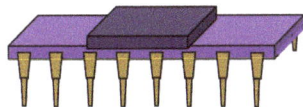

So here we have a non-linear device that can change its resistance by means of memory! These are getting towards the Random Access Memory devices or RAM.

So we can actually physically programme the device as to the value of resistance that we require and that value will be retained in all conditions, with or without power on the unit.

Linear Resistors:

Resistor Colours:

On smaller wattage Resistors, to identify the resistance the manufacturers use coloured bands. I have put the common standard of just three bands in the diagrams but sometimes there can be four.

On larger Resistors the colour bands are not on them because they will have an individual set value which may be printed onto the Resistor.

Colour:	1st No.	2nd No.	3rd No.	Multiply by:
Black	0	0	0	1
Brown	1	1	1	10
Red	2	2	2	100
Orange	3	3	3	1K
Yellow	4	4	4	10K
Green	5	5	5	100K
Blue	6	6	6	1M
Violet	7	7	7	10M
Grey	8	8	8	100M
White	9	9	9	1G

The first four coloured bands are to do with resistance, so if a Resistor only had three bands which were Brown, Blue and Orange. Brown = 1, Blue = 6 and the third is the multiplier, so Orange here is the multiplier (x 000) so the resistance would be 16000Ω or $16K\Omega$.

The band on the other end of the Resistor is known as 'Tolerance' and these can be many different colours as below. These might determine the accuracy of the resistance value.

Remember the value colours by made up sayings, for example those shown in yellow below:

Black	Brown	Red	Orange	Yellow	Green	Blue	Violet	Grey	White
Big	Boys	Race	Our	Young	Girls	But	Vicky	Generally	Wins
0	1	2	3	4	5	6	7	8	9

Tolerance Bands:

Resistors may vary very slightly from their nominal value on the chart below so there would be a 'Tolerance' ring. Resistors are measured for their resistance by the manufacturers at a nominal temperature of 25ºC.

1) Brown +/-1% **2)** Red +/-2%   **3)** Green +/-0.5%   **4)** Blue +/-0.25% **5)** Violet +/-0.1%   **6)** Grey +/-0.05%
7) Gold +/-5% **8)** Silver +/-10%   **9)** No Tolerance Band +/-20%

1) Metal Film Resistor (Linear Resistor):

There are many types of Fixed and Linear Resistors, the one in the diagram below is a **'Fixed' Metal Film Resistor** and one of the most common!

I am sure that you come across many of these Resistors in your world of electronics. They are axial Resistors i.e. Electrode out of either end.

The reason they are called 'fixed' is because the resistance, indicated by the coloured bands, does not change. Inside the resistor there is a coil around a ceramic carrier made up from a thin metal film, hence the name.

The length of the thin metal film, usually Nickel Chromium, determines the resistance, the whole internal coil being covered with a protective coating.

2) Carbon Composition Resistor (Linear Resistor):

There are many types of Fixed and Linear Resistors, the Carbon Composition Resistor in the diagram below may look like the Metal Film Resistor but have a different internal composition.

These Carbon Composition Resistors were abundant in the 1960s, but today they are not used so much.

The Resistor interior is a cylinder with a mix of powdered Carbon or Graphite and clay. The outer protective covering is usually plastic.

As mentioned above these are a fixed resistance and the value of Ohms depends on the length of the cylinder and the amount of Carbon content.

3) Carbon Film Resistor (Linear Resistor):

Carbon Film Resistors are a different improvement on Carbon Composition Resistors. Interiors of both are Carbon Based. Again these are earlier than their Metal Film Counterparts.

So here again there is an Axial Resistor with a ceramic base and pure Carbon spiralled around it inside of a Silicon coating.

This resistor can stand higher temperatures than the Carbon Composition Resistor. There are many types of Carbon Film Resistors some examples are: Leaded, Zero Ohm, Through Hole, etc.

4) Carbon Pile Resistors (Linear Resistor):

These Resistors are made from several Carbon Discs on top of each other compressed between two plates in an outer frame.

Different again from Carbon Composition Resistors or Carbon Film Resistors. Used in many household items as a speed control. These Resistors can also be used in load testers for batteries such as a car battery.

5) Thick Film Resistor (Linear Resistor):

Again these Thick Film Resistors are very similar in looks to the ones we have already discussed, It is just the interior make up that makes them different.

These are axial (electrode at either end) fixed Resistors so their resistance is clamped by the manufacturers.

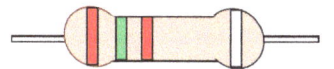

So we start of here with a ceramic tubular shape, and the Resistor part is a thick paste containing Aluminium Oxide Powder, Silver, an inorganic material and a type of glass soldering called glass frit.

The film thickness here is many times thicker than a Thin Film Resistor.

6) Thin Film Resistor (Linear Resistor):

Thin Film Resistors are made exactly the same as Thick Film Resistors except that the materials used are different and the film layer, as per its title is very thin indeed, somewhere in the region of 0.1 Microns!

Thin Film Resistors seem more accurate and preferable.

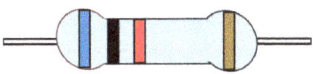

If you had a Thick Film Resistor alongside a Thin Film Resistor you may not be able to tell them apart. Only distinguished by manufacturer's info.

Construction Materials of the film, although other materials may be used, perhaps Lead Oxide or Nickel Chromium (Nichrome). These materials might not be that environmentally friendly.

7) Metal Oxide Film Resistor (Linear Resistor):

Metal Oxide Film Resistors seem to be a bit more complex in construction than other 'Film' Resistors. Remember this is a Fixed Linear Resistor not the Non-Linear Metal Oxide Varistor.

Again, very similar in looks to other 'Fixed Film' Resistors. They actually replaced Carbon based Resistors. Less common than other Resistors.

Just as in other 'Film' Resistors the material film is inserted onto a ceramic base. The Metal Oxide might be Tin Oxide. As above, fixing the Metal Oxide to the ceramics seems more complex than other Fixed Film Resistors.

8) Cermet Film Resistors (Linear Resistor):

These Cermet Film Resistors, (Cermet being more of a Trade Name for a range of Resistors), get their name from the interior material i.e. **Cer**amic **Met**al. The manufacture, like other 'Film' Resistors, uses a Ceramic Base.

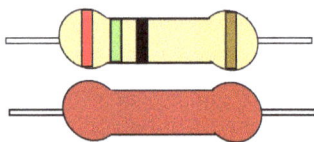

This resistor can be manufactured as Thick Film or Thin Film, the material being different from previous 'Film' Resistors. They come in a range of designs and colours and are not necessarily of the designs in the diagrams on the left. Also some designs are not necessarily axial.

These Cermet Resistors are very heat and flame proof and are designed for high current use. A paste of a mix of ceramic (to tackle heat) and metal (maybe Nickel or Cobalt) is laid onto a ceramic base.

9) Wire Wound Resistor (Linear Resistor):

Before 'Film' and Carbon Resistors were invented this was the main type of Resistor. A coil of resistance wire at a pre-determined length.

10) Tapped Resistor (Linear Resistor):

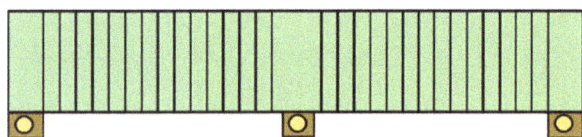

A tapped Resistor is very similar to the Wire Wound Resistor except that there are 'Taps' taken off the coil giving a selection of resistances instead of just one.

11) Fusible Resistor (Linear Resistor):

Although the Fusible Resistor is an actual Resistor it is also a Fuse which protects the circuit against overcurrent and hence it is sometimes called a Current Limiting Resistor. Like a fuse, the interior will melt if an overcurrent flows.

Again we have a ceramic core with resistance wire coiled around it. Can be a 'Film' element or a wound coil in a protective case.

The case may be ceramic as we do not want it to melt or catch fire when the component is carrying out its function.

12) Potentiometer:

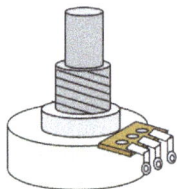

A Potentiometer (Pot) is a Variable Resistance. What we have here is a circular coil of resistance wire. On the right of the diagram are three connections and by connecting wires to them, by means of what is called a 'Wiper' we can, by turning the knob on the top, determine how much of the resistance coil is in the circuit at any one time. There is a range of designs of this component.

13) Rheostat:

A Rheostat, like a Potentiometer, is a Variable Resistance. The one in the diagram to the right might be the most familiar, but they can be round. We could say that a Potentiometer varies the voltage; a Rheostat, again by means of a wiper, uses resistance to vary the current. The coil is made from a Copper Nickel Alloy.

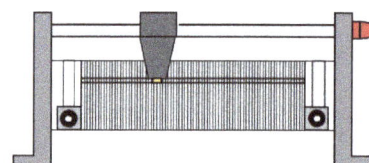

14) Trimmer Resistor:

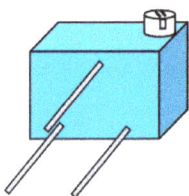

These are sometimes called 'Pots' and, looking at the diagram to the left, the resistance is altered or trimmed by the screw in the top right corner. These can also be obtained as a Variable Cermet Resistor although this may be a trade name. It consists of a resistive element and a wiper that in this case may be called a 'slider', and works similarly to a Potentiometer.

15) Grid Resistors (Linear Resistor)

If you enter Grid Resistors into your search engine you will find that they come in many different forms, one of which I have shown in the diagram below:

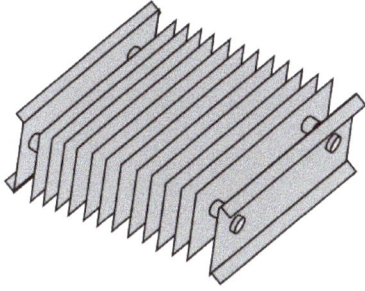

These Resistors can be an array or bank of plates like the example in the diagrams to the left. This is called a bank Resistor because of the number of insulated plates in its makeup.

It will take a very large current and the power may be around 200Watts. These larger Resistors will be used on large amperage equipment such as the railway, ships or large cranes etc.

They have a very good temperature coefficient.

16) Ammeter Shunt (Linear Resistor):

An Ammeter Shunt is very similar to a Varistor as it is connected in parallel to the instrument and in this case an Ammeter. The difference being that the Shunt provides a low resistance path in normality, not just for fault current.

So we connect this low resistance device in parallel with our Ammeter so a large current will pass through the shunt and a small current through the instrument.

By using this device we can use a much smaller Ammeter for a larger current. They can be used for both AC & DC. The Shunt can be made similarly to, and of the same materials used in other Resistors i.e. Wire Wound, Metal Alloy, Ceramic and Carbon.

Unusual Resistors:

1) Liquid Resistors (Non-Linear Resistor):

There are Electric Motor Starters that use liquid as a Rotor Resistance Starter. Although modern technology will make the starters entirely different, the basic technology has been around for years.

When I worked at BP in the 1980s we had Hydromatic Liquid Starters then.

These Electric Motor Rotor Resistance Starters can be called by the name of **Liquid Rheostats.**

2) Aluminium Shell Resistor (Linear Resistor):

As the name suggests these Resistors are made of Aluminium and are high power, up to 200Watts, but compact. One of the other advantages of this Resistor is that it is heat resistant.

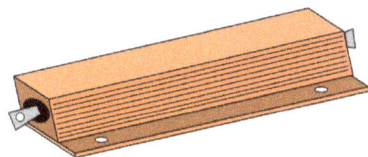

Here we have a Resistor that can be from 1 - 1000Ω (1KΩ). They come in all shapes, sizes and colours. Usually a wire wound Resistor.

3) Braking Resistor (Linear Resistor):

These Resistors may be used on a **Dynamic Braking System**. Can be used anywhere as a brake when kinetic energy to electrical energy is involved.

As one example, we can slow down a DC electric motor by transforming its kinetic (running) energy back into electrical energy. A DC Motor when disconnected from the supply will turn into a generator on spin down, the energy of which can be dissipated through a Braking Resistor. This again is a low resistance, wire wound Resistor.

4) Drain Resistors (Linear Resistors):

Drain Resistors can be put on Capacitors or Batteries to drain them of their charge. Larger wattage Resistors may be recommended here and maybe Wire Wound Resistors.

Resistor Calculations:

Depending upon what you are using your Resistor for, you may have to calculate the value of resistance required.

Resistors can be in 'Series' or 'Parallel' with other components. If we talk about an ammeter Shunt which was discussed earlier then this would be in Parallel with the ammeter.

Series:

If Resistors are in **'Series'** with other Resistors how do we calculate the Resistance? Well firstly the current through each Resistor is the same.

Resistors connected in Series is the easiest:

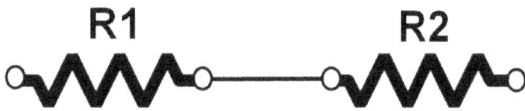

R1   **R2**

Resistors in Series is just a case of adding up the values. So if in the diagram to the left R1 = 6Ω and R2 = 4Ω then 6Ω + 4Ω = 10Ω

Parallel (Two Resistors):

Now we will look at Resistors in **'Parallel'** which is not quite so simple to calculate. The voltage would be the same across each resistor.

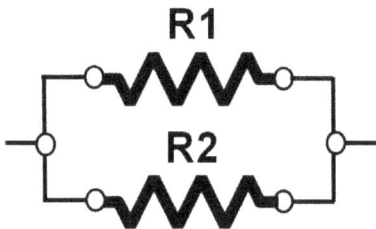

R1

R2

When we calculate the total resistance of two Resistors in parallel as per the diagram to the left using Resistor values of R1 = 6Ω and R2 = 4Ω the formula is:

$$\text{R Total} = \frac{\text{R1} \times \text{R2}}{\text{R1} + \text{R2}} = \frac{6\Omega \times 4\Omega}{6\Omega + 4\Omega} = \frac{24\Omega}{10\Omega} = 2.4\Omega$$

Parallel (Three or more Resistors):

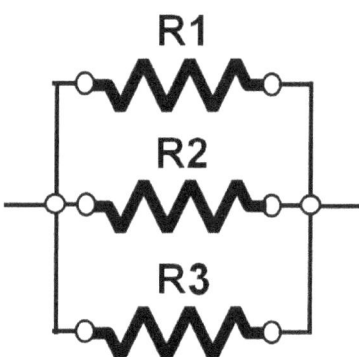

R1

R2

R3

The formula below can also apply if there are only two Resistors.

When we calculate the total resistance of three or more resistors in parallel as per the diagram to the left, using Resistor values of R1 = 6Ω, R2 = 4Ω and R3 = 5Ω, the formula is:

$$\frac{1}{\text{R Total}} = \frac{1}{\text{R1}} + \frac{1}{\text{R2}} + \frac{1}{\text{R3}} = \qquad \frac{1}{\text{R Total}} = \frac{1}{6\Omega} + \frac{1}{4\Omega} + \frac{1}{5\Omega}$$

$$0.17\Omega + 0.25\Omega + 0.2\Omega = 0.62\Omega \qquad \frac{1}{\text{R Total}} = \frac{1}{0.62\Omega} = 1.6\Omega$$

Try a test with both formulas using the same resistance values and you should come up with the same answer!

Testing Resistors:

Testing Resistors is the easiest of all electronic components. The multi-meter is set to the correct scale in Ohms and put straight across the Resistor.

It is important to remember not to get your fingers in contact with the metal of the probe whilst carrying out the test otherwise, in the lower scale, your body resistance may be included in the resistance value!

Remember that although the diagrams below have only three bands, the third being the number of zeros, there could be four bands, the last being the number of zeros (Multiplier).

Resistance Band Values:

To remember the values of the resistors, as suggested earlier, make up a saying that includes all of the letters in the Resistor colours; it does not matter what it is.

Big Boys Race Our Young Girls But Vicky Generally Wins

Black	Brown	Red	Orange	Yellow	Green	Blue	Violet	Grey	White
Big	Boys	Race	Our	Young	Girls	But	Vicky	Generally	Wins
0	1	2	3	4	5	6	7	8	9

Tolerance Bands:

Resistors may vary very slightly from their nominal value on the chart below so there would be a 'Tolerance' ring. Resistors are measured for their resistance by the manufacturers at a nominal temperature of 25°C.

1) Brown +/-1% **2)** Red +/-2% **3)** Green +/-0.5% **4)** Blue +/-0.25% **5)** Violet +/-0.1% **6)** Grey +/-0.05% **7)** Gold +/-5% **8)** Silver +/-10% **9)** No Tolerance Band +/-20%

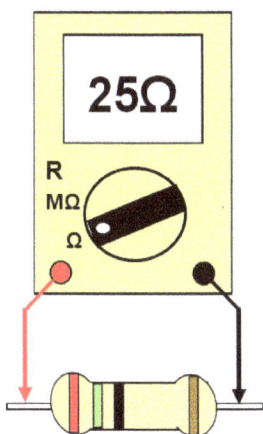

The diagram to the left shows a Resistor that has the coloured bands: Red, Green & Black. So look at the diagram above. The first two colours are the first two numbers of the value and the third colour is the number of zeros on the end. The band on the end is the tolerance.

Red = **2**, Green = **5** so Total = **25** and the black being the number of '0's = **0**

So the value of the Resistor to the left is **25Ω**.

The tolerance band to the right hand end of the Resistor in this case is Gold = +/- 5% so 1% of 25 = 0.25% so 5% = 0.25 x 5 = 1.25% therefore the value could be from **23.75Ω (-5%) to 26.25Ω (+5%)**.

The diagram to the right shows a Resistor that has the coloured bands: Blue, Black & Red. So look at the diagram above. The first two colours are the first two numbers of the value and the third colour is the number of zeros on the end. The band on the end is the tolerance.

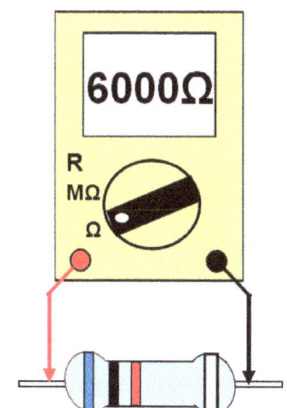

Blue = **6**, Black = **0** so Total = **60** and the red being the number of '0's = **2**

So the value of the Resistor to the left is **6000Ω. (6kΩ)**

The tolerance band to the right hand end of the Resistor in this case is Silver = +/- 10% so 1% of 6000 = 50% so 10% = 60 x 10 = 600% therefore the value could be from **5400Ω (-10%) to 6600Ω (+10%)**.

Resistor Questions and Answers:

Q1 – Is a Resistor a passive device?

A1 – **YES.** It uses energy, but does not produce it like a battery does.

Q2 – What is the difference between Linear Resistors and Non-Linear Resistors?

A2 – **STABILITY.** Linear Resistors are stable where the resistance will not change whereas in Non-Linear Resistors the resistance can vary.

Q3 – What do Resistors actually do?

A3 – **STEM THE FLOW OF ELECTRONS.**

Q4 – Can I test a Resistor using a Multi-meter?

A4 – **YES.** This is one of the easiest components to test. Just switch the meter selector to Ohms (Ω) and connect the test leads across the Resistor.

Q5 – Are Resistors 'through hole' fitting?

A5 – **MOSTLY.** Although surface mounted Resistors can be obtained.

Q6 – Do resistance values have a tolerance?

A6 – **YES.** Most electronic components no matter what they are have a tolerance on their value. Usually the more expensive the component the smaller the tolerance.

Q7 – Can a Resistor be liquid?

A7 – **YES.** There are motor Rotor Resistance Starters that use liquid. They use artificial impurities in the water to create a varying resistance. These days they may be called Liquid Rheostats.

Q8 – What are 'Drain' Resistors?

A8 – **CAPACITORS.** Some circuits where Capacitors are involved use 'Drain' Resistors to dissipate the charge.

Q9 – Does a Resistor hold charge?

A9 – **NO.** Quite the opposite really, but other components in the circuit may.

Q10 – Is a Variable Resistor linear?

A10 – **YES.** I am sure some will debate this with me by the word 'variable' but whichever resistance you set the Variable Resistance to, then it is stable at that setting.

Q11 – Is there a component called a 'Fusible' Resistor?

A11 – **YES.** This component is a Fuse as well as looking like a Linear Resistor.

Q12 – Are there still Wire Wound Resistors?

A12 – **YES.** Although modern Carbon Resistors have taken their place in a lot of cases. Very large current Resistors may still be Wire Wound Resistors.

Q13 – Can Resistors explode?

A13 – **NOT USUALLY.** It would be very unusual for a Resistor to explode. They usually just fail open circuit or burn. This usually occurs if the Resistor is running close to its maximum.

Inductor Section:

An Inductor, solenoid, toroid, choke, ballast, reactor or just plain coil is a passive component, which means it uses energy rather than produces it.

Inductors can be used to filter out low and high frequency noise. There is a rule: Low inductance on High Frequency and High Inductance on Low Frequency.

Solenoids are usually straight coils used in, say, a Relay or a Contactor. Chokes and Ballasts are used in fluorescent lighting.

This component can be used to store energy, similar to a Capacitor, but in a slightly different way, and release it into a circuit at a designated rate and time. A Capacitor stores its current on plates as we have discussed in an earlier section, whereas the Inductor stores its current in a magnetic field.

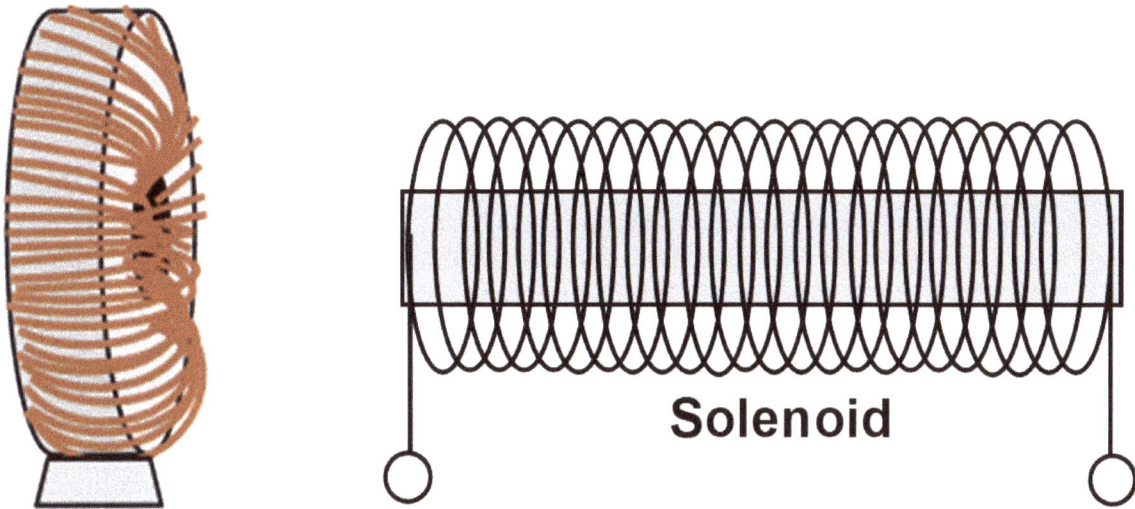

Solenoid

Inductor Cores can be different materials. Four materials for example can be 1) Air, 2) Iron, 3) Ferrite & 4) Ceramic.

Ferrite & Iron Cores tends to be the most preferable in Inductors because of the energy capacity and size of magnetic field.

Let us have a look in this section at what an Inductor actually is, how it works and how we might test it.

NOTE:

ON ALL ATEX EQUIPMENT FOR HAZARDOUS AREAS SUCH AS ZONES, REMOVAL/CHANGING OF INDUCTORS FROM A MANUFACTURERS' PRINTED CIRCUIT BOARD CAN ONLY BE CARRIED OUT WITH MANUFACTURER APPROVAL IN WRITING.

FAILURE TO COMPLY WITH THIS MAY BE CLASSED AS AN UNAUTHORISED MODIFICATION AND MAY RENDER ANY CERTIFICATION DOCUMENTATION NULL AND VOID.

ALSO THE DESIGN, MANUFACTURER, TYPE ETC., OF THE INDUCTOR COMPONENTS MUST NOT BE CHANGED IN ANY WAY.

The Inductor:

What are Inductors?

The Inductor, Toroid or Coil on a printed circuit board firstly, like many of the other components we have talked about, is a passive component, which means that they use energy and do not produce it. An Inductor is very similar to a Capacitor in that it stores power, but in a different way.

Used quite a lot in Radio and TV or where an AC filter or storage may be required. There are two terms which can be attributed to coils that are slightly different: 1) Solenoid: is a straight coil, and 2) Toroid: is a circular coil like the one in the diagram below where the magnetic field created by the coil shape will be much stronger.

The Inductor:

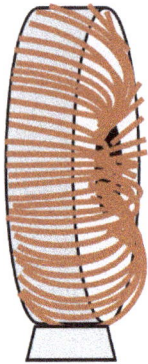

Other names for Inductors are Chokes, Ballasts, Coils or Reactors. The task of the Inductor is to manage current surges by, as above, like a Capacitor storing the surge in its electromagnetic field and releasing that power, when convenient, back into the circuit. We could say that the Inductor is a type of surge filter.

So what are Inductors made of? Well, like a Transformer, there is firstly a core that can be other shapes than the diagram on the left. The core is usually made of Iron or Ferrite and it will generate a large magnetic field. However, I should just mention that the coil does not necessarily need a core.

How do Inductors work?

So how do these Inductors work for the circuit? A Capacitor stores energy on its plates; an Inductor stores energy in a magnetic field. When the power is switched on it takes time for the coil to reach saturation point, in the meantime the magnetic field begins to grow in strength. (Faraday's Law.)

The magnetic field emanates from the core and the direction of the field depends upon the direction of the current flowing through the coil. You can use the right hand thumb rule. It is possible to obtain Inductors that are polarity conscious. The magnetic field of a Solenoid is very similar to a bar magnet and protrudes outside of the coil, but with a Toroidal Coil the magnetic field remains within the core because of the closed loop.

Current through a coil creates a magnetic field: Faraday's Law! If the power to the coil is interrupted in any way then the electromagnetic field will collapse and turn back into electronic energy.

What are inductor measuring units?

The 'storing' of energy in the magnetic field is called 'Inductance' so what values will the Inductor be measured in? The answer is 'Henrys' after Joseph Henry i.e. Micro-Henrys (μH), Milli-Henrys (mH) or Henrys (H). These components might be obvious coils that look like the diagram above, or more subtle.

Measurement depends upon certain factors such as: 1) The length of the wire making up the coil, 2) How many turns of wire are there in the coil, 3) The cross sectional area of the coil, 4) The core material. 5) Permeability.

What can go wrong with an Inductor?

The main thing that can go wrong of course is if, for some reason, the coil is open circuit. Using a multi-meter the reading set on continuity would be infinity. Check over the coil with a magnifying glass and see if you can see an open circuit. Join the leads of the meter together to ensure that you have continuity and then try again.

The second thing that may go wrong is that the coil could short out. This is much more difficult to find with a multi-meter as what you are measuring is the resistance of the wire plus the leads of the multi-meter. So the meter reading on a short circuit may be extremely low and the meter may not read accurately enough and besides you would have to know the resistance of the Toroid to start with.

Fault Finding & Testing an Inductor:

When we look at Inductors on a Printed Circuit Board there is usually a very limited amount of faults that can go wrong with it. Usually the coil is shorting or has gone open circuit.

It is always handy to have a new coil available so that test results can be compared against what they should be!

Remember that Certified Equipment such as Atex must not in any way be bodge repaired or modified and I cannot stress this too much, because if you carry out an unauthorised modification on this equipment Manufacturer's Certificates of Conformity etc. will be rendered null and void!

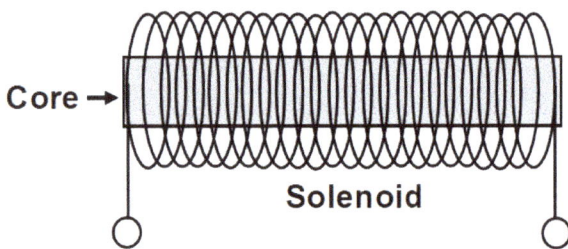

Core →

Solenoid

I have drawn our Inductor (left) as a straight coil for ease of explanation. There are 'X' number of coil turns and a central core.

We have a fault on this component where it will not operate correctly so how are we going to go about finding the problem? Let us look at possible faults and how to test:

Coils Shorting Out:

Finding this problem would not be as easy as it sounds.

Putting a Multi-meter onto the coil as in the diagram on the right would not achieve too much in my opinion as you would get a reading in Ohms (Ω) faulty or not. **You would need to know the resistance of a good coil first then check the resistance against this one.**

The fault may be obvious so get out the magnifying glass and have a close look for maybe burned coil turns?

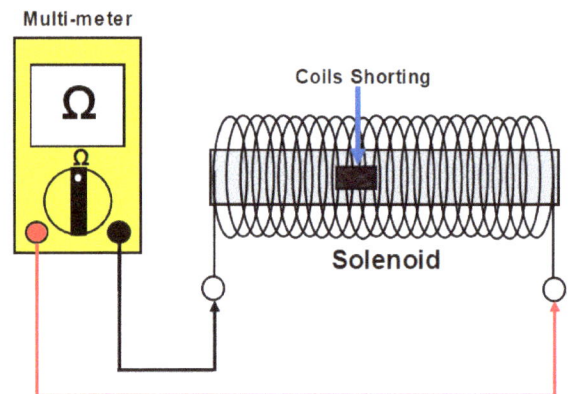

Multi-meter

Coils Shorting

Solenoid

Coils Open Circuit:

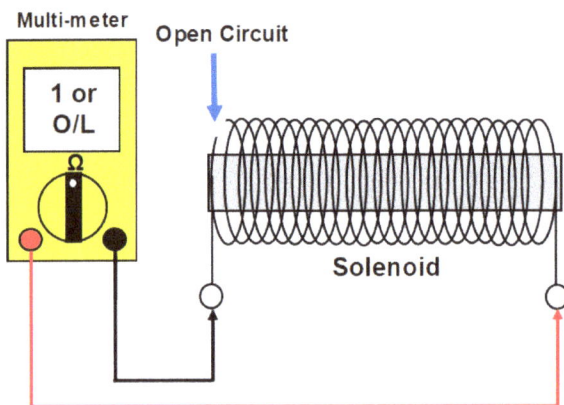

Multi-meter Open Circuit

1 or O/L

Solenoid

Finding this problem would not be easy either.

All you are doing here really is checking if this coil is the cause of whatever is going wrong with your circuit. Move the probe along the coil like a jockey.

By putting the Multi-meter onto the coil as the diagram to the left, all you are doing is checking if there is continuity through the coil. If there is you will get a reading in Ohms (Ω), if not you will get the reading of 1 or O/L.

Either of the faults above would require a new component! If the equipment is Atex then any bodge repairs would be out of the question.

All the way through I urge technicians to get out a magnifying glass and see if they can see anything wrong before going through a lot of complicated testing, most of which would have to be carried out by removing the components from the circuit.

Inductors:

Types of Inductor & their uses:

When looking at the different types of inductor below take particular notice of the material that the core is made up from as this is the basis of the Inductor Characteristics.

Different materials determine the frequencies that the inductor can be used in, for instance a solid Iron core can only be used at low frequency, going to a powdered core which will go up to much higher frequency, and at the top of the frequency band, an Air Core.

Remember that Coils/Inductors are found in equipment such as Printed Circuits, LAN Systems, SEPIC Systems, Radio, Lighting, Relays, Contactors, Motors, Current Transformers, Overloads, Electromagnets and many other items of equipment.

Let us look at thirty common and not so common inductors; I am sure that there are many more:

1) **Ferrite Core Inductor:** Passive component with Ferrite core within the coil and a closed loop of current. Suitable for high frequency applications and interference suppression. Broadband comes to mind as one of the uses.

2) **Toroidal Inductor:** Can be commonly found on printed circuits. Looks like a ring of wire. Diagram on previous page. Used in Telecommunications and Medical Instruments amongst other equipment.

3) **Axial (Coloured Ring) Inductor:** Very small coil that looks very similar to a Resistor. Used in Filters, ideal for Printed Circuits.

4) **Bobbin Inductor:** Looks like a cotton bobbin with a Silicon Steel laminated core. Used on Switch Mode Power Supplies (SMPS) and Filters.

5) **Wireless Charging Coil:** Used primarily for inductive charging on equipment such as Communication Systems, smartphones, laptops.

6) **Multi-layered Chip Inductor:** Smaller than the common wire wound inductors. Oblong in shape and uses include Bluetooth & LAN (Local Area Network) Wi-Fi etc.

7) **Coupled Inductor:** Two separate windings on an encased core. Used in SEPIC (Single Ended Primary Indictor Converter).

8) **Shielded Surface Mount Inductor:** Small and compact, ideal for printed circuits. Used in PoL (Point of Load) Converters & anything where batteries are involved.

9) **Shielded Variable Inductor:** Movable core inside of the coil winding gives the effect of changing the inductance. Used in Automotive applications.

10) **Air Core Inductor:** The coil is formed by wrapping around a small cylindrical tube. No core such as Ferrite, as it uses only air as the core. Used as Radio Frequency (RF) Tuning Coils.

11) **Laminated Core Inductor:** Very Common. Low eddy currents due to the laminated core. Used in Power Filtering Systems.

12) **Powdered Iron Core Inductors:** These Iron particles are compressed and mixed with epoxy to form the core. More magnetic flux than a ferrite core. Used in Dimmer & Filter Chokes etc.

13) **Audio Frequency Inductors (Chokes):** Suitable for use on low frequency tasks.

14) **Radio Frequency Inductors (Chokes):** Wire is wound onto an air coil. Suitable for high frequency tasks.

15) **Moulded Inductor:** Coated with an insulation material such as moulded plastic with a ferrite core. Very small so can easily be used on a printed circuit board.

16) **Coupled Inductor:** Works a bit like a transformer with two windings. One winding induces into the other. Can be used in SEPIC (Single Ended Primary Indictor Converter).

17) **Thin Film Inductor:** Internal spiral magnetic material coils formed on a substrate for the induction process to work. Used in communication devices, wireless systems and LAN (Local Area Network) systems.

18) **Ferromagnetic/Iron Core Inductor:** Ferromagnetic core increases the induction. Limited application. Audio Systems.

19) **Choke:** There are many types of choke, some are mentioned earlier. Possibly the most popular chokes are used in fluorescent lighting. Chokes are usually used to limit the current.

20) **Ballast:** Again more commonly known for their use in fluorescent lighting. Not quite the same as a Choke. A Choke tends to be very inductive whereas a Ballast can be resistive as well as inductive.

21) **Planar Inductors:** High powered inductors use what is called a flat wire winding. High power applications. Used in RFICs (Radio Frequency Integrated Circuits).

22) **Ceramic Core inductors:** The core would be non-magnetic so just like an air core and less core losses. Used in low inductance applications,

23) **Roller (Coaster) Variable Inductor:** Very similar to a Rheostat/Auto-Transformer in the fact that the coil length can be varied. The inductance varies according to the coil length.

24) **Bifilar Inductor:** Consists of two parallel coil windings enabling a greater magnetic field.

25) **Ferrosilicon (FeSi) Core Inductor:** These are Iron (Fe) and Silicon (Si) Core Inductors. Can be chokes and can store a lot of energy in their powerful magnetic field.

26) **Shielded Inductor:** This inductor may be useful on printed circuits as it contains its magnetic field inside of its enclosure so that surrounding components are not affected.

27) **Wire Chip Inductor:** These are usually printed circuit 'surface mount' inductors as opposed to 'through hole' inductors. Used in radio/wireless situations.

28) **Binary Switched Inductors:** Two coils in series doubles the inductance. Coils can be shorted out to create a variable inductance. Not quite the same as a Variometer.

29) **Variometer:** There are two variations of Variometer, one is concerned with aircraft and one is an inductor where two coaxial coils are connected in series. Their axis position relative to each other can be altered for tuning a circuit.

30) **Slug Tuned Inductor:** The Ferrite core moves in and out of the coil, varying the inductance. When the core slug moves into the coil the inductance will increase. Uses are Radio, Power Factor Correction, and high frequency uses amongst many other things.

Inductor Symbols:

Most Common Inductor Symbols:

Above are, possibly, the two most common symbols for inductors on a schematic diagram. These are usually fixed value and possibly the type such as Air Core Resistors.

Other Common Inductor Symbols:

Below are several more common symbols that you might see:

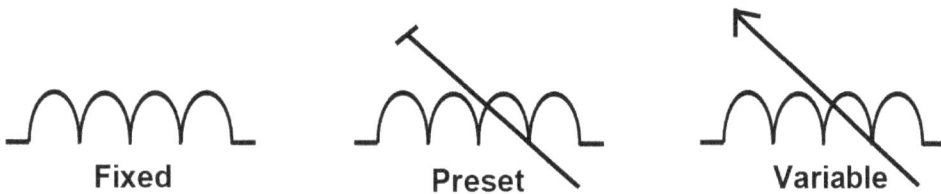

Fixed   **Preset**   **Variable**

Air Core Inductor

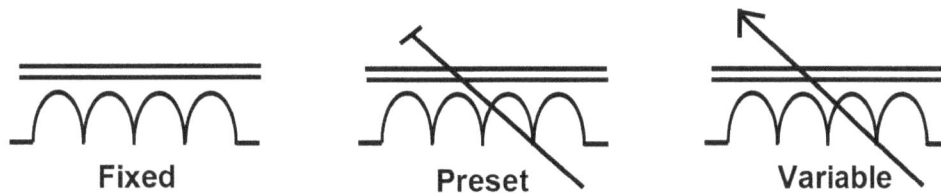

Fixed   **Preset**   **Variable**

Iron Core Inductor

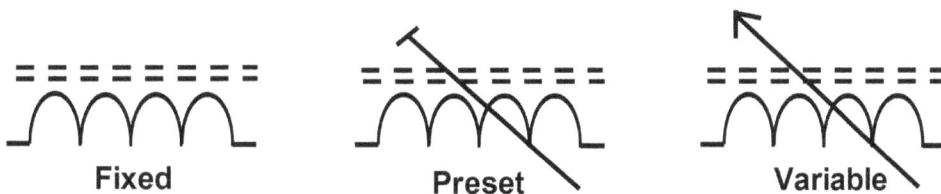

Fixed   **Preset**   **Variable**

Ferrite Core Inductor

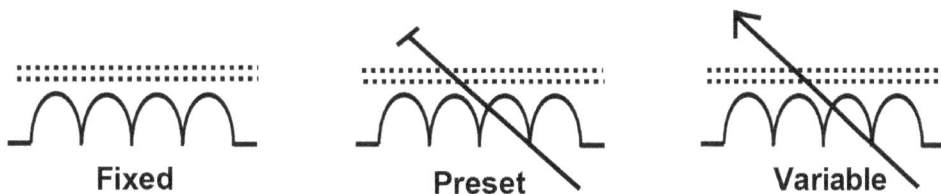

Fixed   **Preset**   **Variable**

Iron Dust Core Inductor:

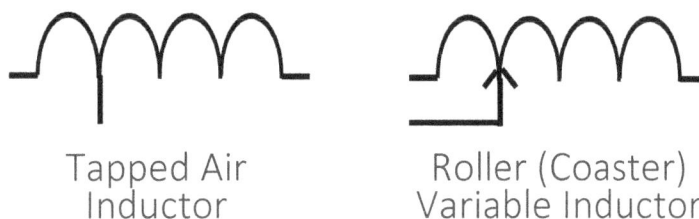

Tapped Air
Inductor

Roller (Coaster)
Variable Inductor

PCB Inductor Questions and Answers:

Some of the questions and facts below may refer to Inductors in general, rather than just PCB Inductors.

Q1 – What is the phenomena of storing energy in a magnetic field called?

A1 – INDUCTANCE.

Q2 – What units is Induction measured in?

A2 – THE HENRY. After Joseph Henry, an American Scientist 1797-1878.

Q3 – What is another name for an Inductor?

A3 – SOLENOID – TOROID – CHOKE – BALLAST – REACTOR

Q4 – What is the purpose of an Inductor?

A4 – THEY STORE ENERGY IN A MAGNETIC FIELD.

Q5 – Do integrated circuits have Inductors?

A5 – NO. Inductors cannot be miniaturised enough.

Q6 – What are the best coil core materials?

A6 – IRON OR FERRITE. Air is possible but not as effective.

Q7 – What are the main faults with Inductors?

A7 – SHORT OR OPEN CIRCUITS. Sometimes if the inductor is short circuiting you will see burning within the coils which usually means a new component. Open circuits are much harder to locate and usually discovered through testing and again, usually means a new component.

Q8 – Does current affect the magnetic field?

A8 – MOST CERTAINLY. The field strength is proportional to the current magnitude.

Q9 – Can inductors go open circuit?

A9 – THEY MOST CERTAINLY CAN. Continuity tests in ohms may not pick up a coil short circuit. Have a close look with a magnifying glass.

Q10 – What is the main drawback with Inductors?

A10 – SIZE AND EFFECT ON OTHER CLOSE COMPONENTS AROUND THEM.

Q11 – What are the units of Inductive Reactance?

A11 – OHMS (Ω)

Q12 – What determines the Inductance of an Inductor?

A12 – NUMBER OF TURNS – PERMEABILITY – COIL LENGTH – COIL AREA.

Q13 – Does the current lead or lag the voltage in an Inductor?

A13 – THE CURRENT WILL LAG THE VOLTAGE. Leads in Capacitance! The size of our PCB Inductor Power Factor will be insignificant.

The Integrated Circuit Section:

Integrated Circuits can be Analogue (i.e. Oscillators), Digital (i.e. Ram & Rom) or a mixture of the two. They can contain many components such as Transistors. Their size makes them more attractive than a Printed Circuit. Probably the only electronic components that they cannot contain would be an Inductor or Toroid and a Transformer, which would be too large.

It has got to such a stage that we would not have home computers, laptops, electrical appliances and the ability to send rockets to the moon without the Integrated Circuit.

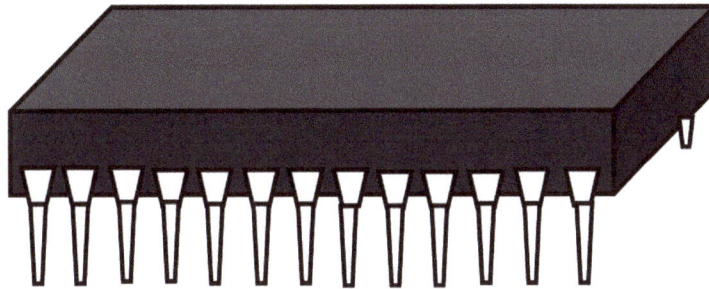

Here we could have millions of Transistors on the surface of a piece of Silicon inside a framework similar to the above design that is no bigger in surface area and volume than an i-phone sim card.

1) A microchip starts with a small area of Silicon called a **'Wafer'** which is a smaller section of a factory grown crystal.
2) The **'Wafer'** is the coated with a material called **'Photoresist'** and when exposed to ultra-violet light this creates a pattern on the wafer called **Masking.**
3) **Etching** then takes place in the **Photoresist Pattern.**
4) Millions of components created on the surface by doping, which we discussed earlier, making millions of PN junctions in atmospheres that are super clean.
5) The process is much more complex than how I am trying to explain it here!

We will discuss Logic Gate **'Truth Tables'** along with Manufacturer's **'Pin Out'** Diagrams for Integrated Circuit Chips.

In this section we will have a look at how the microchip carries out its role in the world of electronics without going into too much depth.

NOTE:

ON ALL ATEX EQUIPMENT FOR HAZARDOUS AREAS SUCH AS ZONES, REMOVAL/CHANGING OF INTEGRQATED CIRCUIT CHIPS FROM A MANUFACTURERS' PRINTED CIRCUIT BOARD CAN ONLY BE CARRIED OUT WITH MANUFACTURER APPROVAL IN WRITING.

FAILURE TO COMPLY WITH THIS MAY BE CLASSED AS AN UNAUTHORISED MODIFICATION AND MAY RENDER ANY CERTIFICATION DOCUMENTATION NULL AND VOID.

ALSO THE DESIGN, MANUFACTURER, TYPE ETC., OF THE INTEGRATED CIRCUIT CHIPS' COMPONENTS MUST NOT BE CHANGED IN ANY WAY.

The Integrated Circuit:

Description:

People ask 'what is an Integrated Circuit (IC)?' Well if you look at the electronic components we have been talking about on printed circuits you will see that although the components seem to be small, in relative terms they are not. A printed circuit full of Resistors, Diodes, Capacitors etc. sometimes looks like a complex cluttered arrangement.

If there was a way to achieve the same objective in miniature and inside of one component then life would seem much less complex. Well now they have done just that; all of the electronic components are inside of a container called an IC Chip, sometimes called a Microchip. A Microchip can be part of the printed circuit.

Microchips these days are in all electronic appliances. They are very reliable, but unlike a printed circuit there are no physical components to check or change. The chip either works or it does not.

Diagram of an Integrated Circuit:

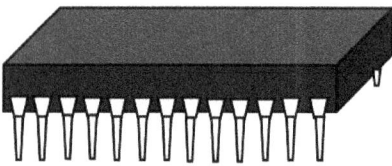

The Microchip can be any shape and size and can have any design of components that is required. I am not going into how the Microchip is made as it is not my field of electronics; all I will say is that the main part of the chip is silicon, with the circuit and components etched on.

We can end up with very small Chips that can house thousands of Transistors. Just imagine what size a normal printed circuit would have to be! There are different types of Microchip depending on what you want it to do. Two types are Logic and Memory, both of which are in a Computer.

Logic Microchips:

Logic Microchips, for instance, have what is called a CPU (Central Processing Unit). If you look at a Computer the CPU is the part that carries out the instructions. Another type is an NPU (Neutral Processing Unit) and these are used, for instance, in self-driving cars.

Memory Microchips:

The Memory Microchip, as the name suggests, just stores information. Even here there are two types; the first is DRAM (Dynamic Random Access Memory). If you take a Computer as an example, this is the memory that disappears when the machine is turned off.

The second is NAND Flash Memory (Not AND). Again, if you compare with a Computer, this saves the information even when the machine is turned off. For example if you are writing a training course on a Computer using Power Point, you do not want the information to disappear the moment that you switch off the Computer.

Motherboard:

Motherboards are integrated circuits. They are the heart of a Computer that takes in both Logic and Memory systems above. Sometimes the board requires a CMOS Battery (Complementary Metal Oxide Semiconductor) and this battery is required to supply the board with a small amount of power to hold the memory when the Computer is switched off. CMOS Batteries last around ten years.

Disadvantages:

There are several disadvantages to the Microchip; take for instance the fact that the Chip has a set number of internal components so if one of them fails then a new Chip is required. Connecting extra components to the outside of the Chip is difficult, if not impossible. The third and last thing I will mention is Inductors, it is not physically possible to manufacture any type of Induction Coil small enough to be inserted onto the Silicon surface of a Microchip. There are other disadvantages such as thermal conductivity etc.

General:

Integrated Circuits, as we have mentioned previously, are in most electronic devices these days and there are many different shapes and designs.

The Integrated Circuits can be broken down as follows:

Analogue: These usually consist of circuits that use variable electronic power signals in the form of waves which carry the data. If I took an electronic circuit that contained independent capacitors, resistors etc. then this would be an Analogue Circuit, so the Integrated Circuit would be in an electronic circuit with such components.

Digital: These Integrated Circuits are different from the above in the fact that they 'Integrate' all of the components and connection wires onto one semiconductor chip. Then 'Gates' are used to transfer the data in the form of logic.

The Digital Integrated System can then be broken down further to

a) Small Scale Integrated Circuits (**SSI**) which is the lowest containing 10 Gate Circuits with up to 100 Components,

b) Medium Scale Integrated Circuits (**MSI**) that contain up to 100 Gate Circuits with up to 1000 Components,

c) Large Scale Integrated Circuits (**LSI**) with over 100 Gate Circuits with up to 10,000 Components,

d) Very Large Scale Integrated Circuits (**VLSI**) with over 10,000 Gate Circuits and up to 1,000,000 Components,

e) Ultra Large Scale Integrated (**ULSI**) with over 100,000 Gate Circuits and up to 10,000,000 Components. So you can see the enormous attraction of these systems!

Hybrid and Monolithic:

Integrated Circuits can also be:

a) **Hybrid** which means that there is, say, a Printed Circuit Board with an '**Analogue**' Integrated Circuit as part of it, mixed with Capacitors, Resistors etc. or

b) **Monolithic** where all the components are combined into one '**Digital**' Integrated Circuit Chip. Let us look at a few that can be found on various drawings and in many manuals.

Single Line (SIL):

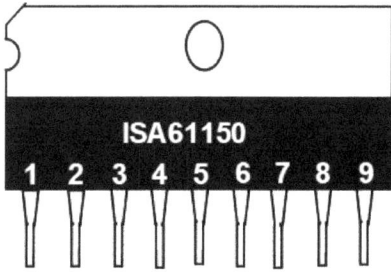

The diagram to the left is what is called a SIL (Single in Line) Integrated Circuit. The pins are numbered from left to right. The section in the diagram above the main chip is a heat sink, which is used to dissipate any heat produced when it is working. So these Integrated Circuits carry out a particular task i.e. an amplifier, in a piece of equipment. They can contain many transistors etc. in the one unit and are cheap and easily changed if things go wrong.

Double Line (DIL):

The diagram to the right is what is called a DIL (Dual in Line) Integrated Circuit performing a particular task. The pins are in two lines numbered in an anticlockwise direction looking down on the chip from above. There is a 'dimple' or a 'Notch' local to pin 1 to avoid confusion as to which way round the integrated circuit is. The actual number of pins relies on the particular duty of the Integrated Circuit.

Schematics:

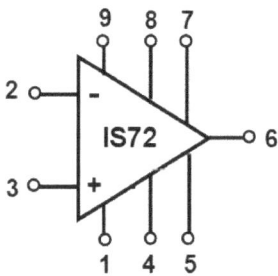

The diagram to the left is how an Operational Amplifier Integrated Circuit performing particular tasks **MAY** be shown on an Electronic Schematic Diagram. Sometimes a diagram of the full Integrated Circuit Chip, in square form, is shown on the drawing. The schematic drawing designer will designate the numbering of the pins on the drawing and add any external electronics. Remember this may be one Integrated Analogue Circuit in the middle of a printed circuit or this could be a good example of a Digital Integrated Chip in an Analogue circuit!

The triangular diagram above may be a small Analogue Integrated Circuit feeding into a much larger Integrated Circuit via other electronic components on a board.

Integrated Circuit Through Hole Chip:

The pins on the Integrated Circuit Chip are a set distance apart which match Vero-board holes or Breadboard holes that you are putting it into. Manufacturers would make their printed circuit board holes fit the Integrated Circuit Chip.

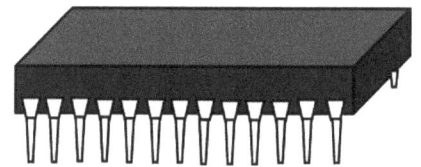

Integrated Circuit Surface Mounted Chips:

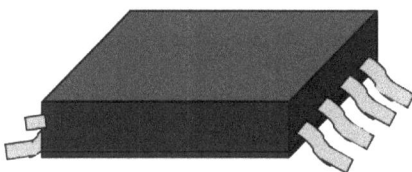

It is possible to get surface mounted Integrated Circuit Chips as the diagram, left. If the IC Chip itself is to be surface mounted the pins are slightly different, so instead of being pointed they have an out-turned flat end as in the diagram left.

Integrated Circuit Socket – DIL Socket:

Sometimes it is not the actual Integrated Circuit Chip that is surface mounted. There is a long device called an Integrated Circuit Socket or DIL Socket that the Integrated Circuit Chip plugs into, and it is this that you surface mount. It is possible that this unit would come in a set with the Integrated Circuit Chip. This device is through hole into the Printed Circuit Board allowing you to plug the Integrated Circuit Chip into it without soldering. The Integrated Circuit Chip can then be removed and replaced easily at a later date. This type of IC Chip would be the type with the pointed through hole pins.

Integrated Circuit ZIF Socket:

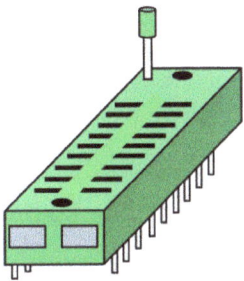

There is another device called a ZIF (Zero Insertion Force) Socket which uses a clamping method to hold onto the Integrated Circuit Pins rather than risk them buckling on the DIL Socket.

These are through hole soldered onto the Printed Circuit Board and again like the DIL Socket, the Integrated Circuit Chip can be removed and replaced without any further soldering. You may find them on a Motherboard where there are a large number of pins from the Integrated Circuit Chip.

Timers:

The next page talks about 'Gates' and how signals are passed through. An Electronic Timer IC Chip would contain many of these 'Gates' connected in a particular sequence.

One of the more common Timer Integrated Circuit Chips is the 555 Timer. It gets its name from the three identical 5kΩ resistors inside of the Chip. If you require an IC Timer this may be the answer.

Logic Gates:

We might count using multiples of numbers from 1 – 10, but computers & calculators etc. do not! They use what is called a Binary Code which they use in a special way. This, to a computer or processor, is a system using **ZEROs & ONEs**. They use, as we have discussed in an earlier section, a Transistor to carry out the switching. It will not switch if the input codes are incorrect. There are many types of Signal/Gate, let us just look firstly at three types: **1) AND; 2) OR; 3) NOT.**

'AND' GATE:

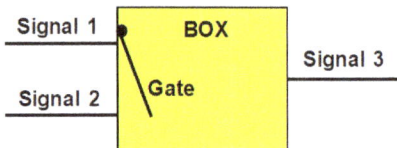

Picture a Box with a Gate, like the one on the left, with two **INPUT** signals (1 & 2), but only one **OUTPUT** signal (3) is allowed through the Gate and out. So our Box has to choose which **OUTPUT** signal that is, using the Binary Code. Look at the water analogy below:

Let us say that the box is an analyser and certain signals are pure water. The **OUTPUT** signal 3 **must be pure water.** Going by this logic the **INPUT** signals 1 & 2 must both be pure water, otherwise if one of the inputs was contaminated the **OUTPUT** signal 3 would not be pure. **BOTH INPUTS MUST ENTER AT THE SAME TIME!** I hope this is not confusing the matter. **So if one input 0 it would not be pure.**

Let us round this up – I will change the shape and name of the Box. Let us now call the Box itself a 'Gate'. The output is either **NO** or **YES** (0 or 1).

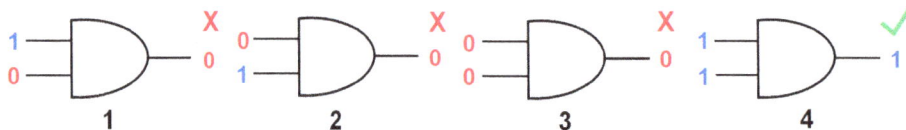

Looking at the above diagram, numbers 1 & 2 input signals are 0 & 1, so going by our water analogy, 0 is not pure so the Gate would not let them past. With diagram 4 the inputs are identical i.e. pure 1 so the output would be the same so the Gate would pass both signals. **It will not open if both signals are** 0 (diagram 3). Look at each diagram (1, 2, 3 & 4) inputs as two switches in series, both to be made for an output.

'OR' GATE:

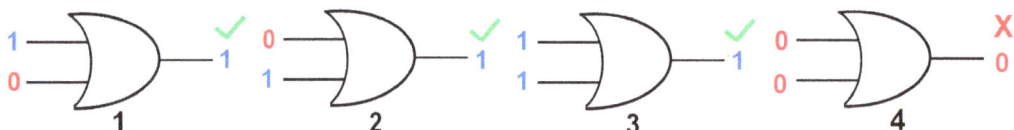

Looking at the above diagram our Box/Gate is a slightly different shape. The number 1 & 2 diagram input signals are still 0 & 1; this time things are slightly different. OR Gate will open if **EITHER** of the **INPUTS** are 1 i.e. 0 OR 1. Diagram 3 input signals are both 1 so the output signal will automatically be 1. **It will not open if both input signals are** 0. Look at each diagram (1, 2, 3 & 4) inputs as two switches in parallel, one switch to be made for an output.

'NOT' GATE:

This Gate is different again to the **'AND'** & **'OR'** Gates. With this one there is only one input signal and one output signal. With a **'NOT'** Gate the output signal is the direct opposite to the input signal i.e. if the input signal is a 0, the output signal will be a 1. If the input signal is a 1 then the output signal will be a 0.

NOR GATE:

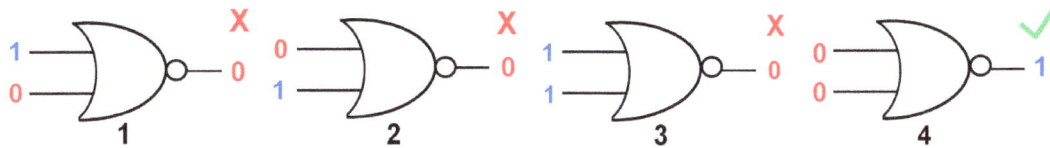

As you can see by the above diagrams the **NOR** Gate differs from the **OR** Gate by a small circle in front of the Gate Logo. What we are basically looking at here is a 'Not OR' Gate, so looking at the above diagram, numbers 1 & 2 input signals are 0 & 1 so in an 'OR' Gate the output would be '1', but as we are 'Not an OR Gate' the output to this Gate would be '0'. Looking at the above diagram 3, the inputs are both '1' so judging by our rule of 'Not an OR' the output here would be '0'. Lastly looking at diagram 4, both inputs are '0' so going by our 'Not OR', the output would be '1'.

NAND GATE:

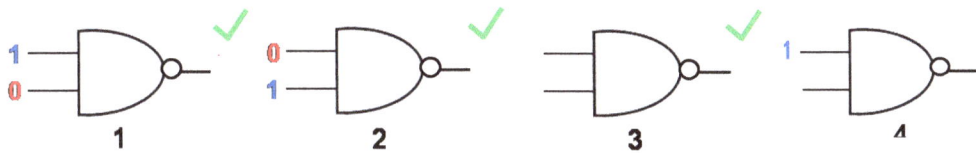

As you can see by the above diagrams the **NAND** Gate differs from the **AND** Gate by a small circle in front of the Gate Logo. What we are basically looking at here is a 'Not AND' Gate, so looking at the above diagram, numbers 1 & 2 input signals are 0 & 1, so in an 'AND' Gate the output would be '0', but as we are 'Not an AND' Gate the output to this Gate would be '1'. Looking at the above diagram 3, the inputs are both '0' so judging by our rule of 'Not an AND' Gate, the output here would be '1'. Lastly in diagram 4, the inputs are both '1' so judging by our 'Not an AND' Gate the output would be '0'.

XOR GATES:

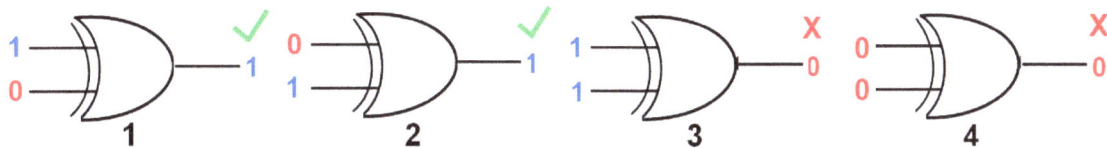

Sometimes these **XOR** Gates would have an 'E' at the start and would be an 'EXOR' Gate, the 'E' or just the 'X' stands for 'EXCLUSIVE' OR Gate. So in this case if we take diagrams 1 & 2, you will see that it is basically the same as an 'OR' Gate where if the input is '1' or '0' then the output being 'OR' would be '1'. The change comes with diagrams 3 & 4. If the inputs are '1' & '1' or '0' & '0' then the output is '0' as the inputs are identical.

XNOR GATE:

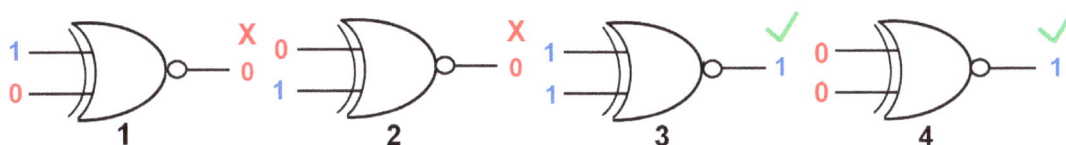

Sometimes these **XNOR** Gates would have an 'E' at the start and would be an 'EXNOR' Gate, the 'E' or just the 'X' stands for 'EXCLUSIVE' NOR Gate. So in this case if we take diagrams 1 & 2, you will see that it is different from an 'OR' Gate where if the input is '1' or '0' the output would be '0', but if the input was either '1' & '1' or '0' & '0' the output would be '1'.

Truth Tables:

Logic Gate:	Gate Inputs: A	B	Gate Output:
AND	1	0	0
	0	1	0
	0	0	0
	1	1	1
OR	1	0	1
	0	1	1
	1	1	1
	0	0	0
NOT	1		0
	0		1
NOR	1	0	0
	0	1	0
	1	1	0
	0	0	1

Logic Gate:	Gate Inputs: A	B	Gate Output:
NAND	1	0	1
	0	1	1
	0	0	1
	1	1	0
XOR	1	0	1
	0	1	1
	1	1	0
	0	0	0
XNOR	1	0	0
	0	1	0
	1	1	1
	0	0	1

When designing or Fault Finding on Integrated Circuits you would need to know what Gate Logic to use or is being used.

Truth Tables are a way of defining what each Logic Gate output is going to be, based on input signals.

I have drawn all of the Logic Gate Systems on one Truth Table, but these are usually formed for each of the seven Logic systems i.e. **AND, OR, NOT, NOR, NAND, XOR and XNOR.**

Truth Tables have other uses, for instance, in mathematics etc.

Pin Out Diagram for Integrated Circuit Chip:

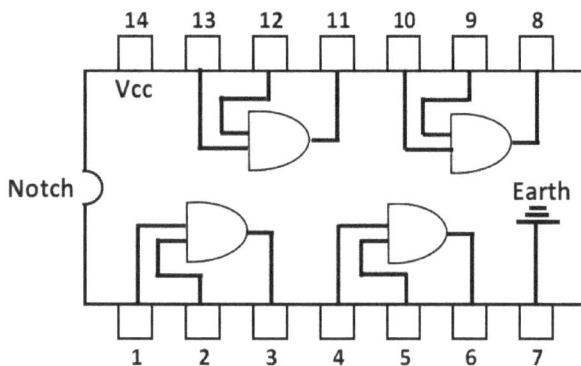

We have discussed Logic Gates and to round things off, looking at the diagram left is what is called a **'Pin Out'** Diagram or **'Data Sheet'** which will accompany the Integrated Circuit Chip to enable correct connecting.

There is the notch if you remember earlier, which determines which way you look at the Integrated Circuit Chip and what the numbering system of the pins is i.e. numbered from 1 in an anti-clockwise direction.

Looking at the above diagram, **Vcc** stands for 'Voltage collector' which will be labelled positive (+) for NPN Gates (Usual), and Negative (-) for PNP Gates. Some Data Sheets may refer to this terminal as **Vdd** or Voltage drain, depending if the component is **BJT** or **FET**. We have discussed this in our Transistor Section.

Usually the **Vcc** will be positive with respect to the Earth which will be **0 Volts** which may be referred to as **Vee**, Voltage emitter (BJD), or **Vss**, Voltage source (FET). So in other words in the case of our diagram above there will be a voltage between **Vcc** and Earth.

So now if you apply what we have discussed in the Logic Gate section to what we have in the above manufacturer's **'Pin Out'** Diagram you may just have an outline of how it all fits in.

Integrated Circuit Questions and Answers:

Q1 – If part of the integrated Circuit fails inside of the chip, can it be repaired?

A1 – NO unfortunately this would have to be a new microchip. The digital gates etc. cannot be repaired.

Q2 – Can an Integrated Circuit Chip be removed from a board?

A2 – YES but not easily. Use temporary heat sinks where possible.

Q3 – How do I select the correct Microchip when renewing it?

A3 – USUALLY BY THE INFORMATION NUMBER ON THE TOP.

Q4 – Can I damage the Microchip with, say, heat from a soldering iron?

A4 – MOST CERTAINLY as above try to use a heat sink.

Q5 – Why would I choose an Integrated Circuit over a Printed Circuit?

A5 – SIZE, LESS SOLDER JOINTS, RELIABILITY etc.

Q6 – Is an Integrated Circuit a low power unit?

A6 – MOST CERTAINLY! They can get very warm and some have built in metal heat sinks.

Q7 – What are Monolithic Integrated Circuits?

A7 – ALL ON ONE SUBSTRATE. All components & circuits are formed on one thin medium.

Q8 – Is a Microprocessor an Integrated Circuit?

A8 – YES they are the same thing.

Q9 – What is ROM?

A9 – READ ONLY MEMORY. This is an Integrated Circuit you will find in your computer that completes certain actions which store memory, but you cannot actually write to it. Let us take an Instruction Manual as an example.

Q10 – What is RAM?

A10 – RANDOM ACCESS MEMORY. Data is stored temporarily on this Integrated Circuit and must be refreshed at intervals and saved before switch off. This is that annoying part where you lose your work if there was to be some kind of power interruption.

Q11 – What are ASICs?

A11 – APPLICATION SPECIFIC INTEGRATED CIRCUITS. They only provide one specific function and have not the scope of their larger counterparts.

Q12 – What are RFICs?

A12 – RADIO FREQUENCY INTEGRATED CIRCUITS. As used in, say, a mobile phone/radio.

Q13 – What is CMOS?

A13 – COMPLEMENTARY METAL OXIDE SEMICONDUCTOR. What the Integrated Circuit is made of. These are so sensitive to static electricity that they can be destroyed, simply by being handled, by the static electricity present in humans!

Oscillator Section:

Oscillators come in all shapes and sizes. They can be a simple electronic component as below, or a larger unit the size of a disc player. They can also be a circuit of components on a PCB. They can be fixed or programmable.

They are, in a sense, a miniature Inverter, but giving an output signal in low power short bursts, unlike an Inverter where the output is constant.

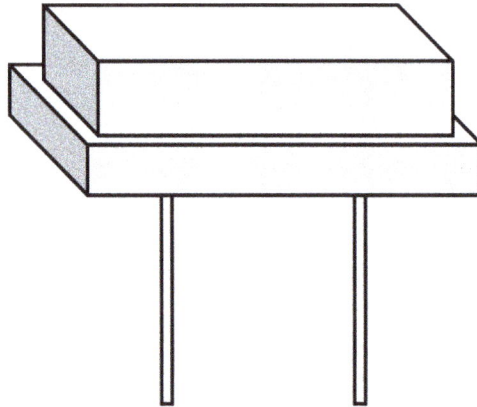

Let us have a look at how an Oscillator works and where it might fit into our Printed Circuit and why we might require an AC pulse.

There are many designs of Oscillator as you will see.

NOTE:

ON ALL ATEX EQUIPMENT FOR HAZARDOUS AREAS SUCH AS ZONES, REMOVAL/CHANGING OF OSCILLATOR FROM A MANUFACTURERS' PRINTED CIRCUIT BOARD CAN ONLY BE CARRIED OUT WITH MANUFACTURER APPROVAL IN WRITING.

FAILURE TO COMPLY WITH THIS MAY BE CLASSED AS AN UNAUTHORISED MODIFICATION AND MAY RENDER ANY CERTIFICATION DOCUMENTATION NULL AND VOID.

ALSO THE DESIGN, MANUFACTURER, TYPE ETC., OF THE OSCILLATOR COMPONENTS MUST NOT BE CHANGED IN ANY WAY.

PCB Oscillator:

The Oscillator Description:

Quite a number of PCBs (Printed Circuit Boards) today have an Oscillator on them. So what is this Oscillator? Well first of all they can be a separate component or part of a microchip.

In this book we talk about the Diode which converts AC (**A**lternating **C**urrent) to DC (**D**irect **C**urrent); this is the easy task. Converting DC to AC is not so easy and on large industrial complexes this is done with a large unit called an Inverter which is fed by huge banks of batteries.

If we want to produce AC from DC on our PCB at a desired uniform waveform it has, obviously, to be done in miniature form and we can do this to a certain degree with a component called an Oscillator.

So to sum up, an Oscillator produces and maintains a waveform at a certain level and frequency. Some of the waveform shapes produced are shown below.

Standalone or Integrated Circuit:

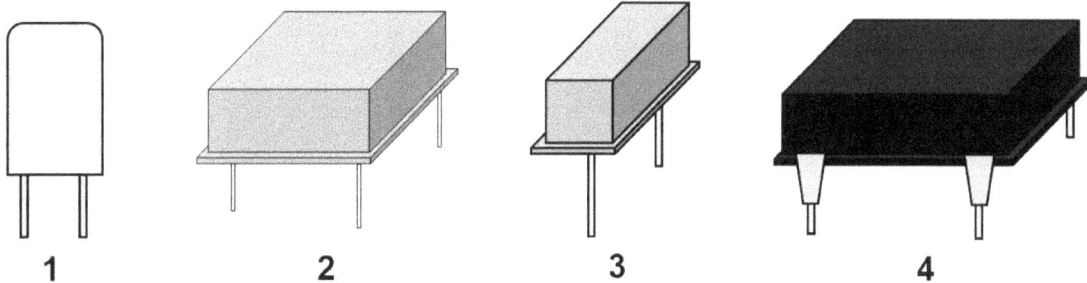

1   **2**   **3**   **4**

Sometimes a PCB Oscillator can be a standalone component like the diagrams 1-3 above, with the circuit inside, or it could be an integrated circuit chip as in diagram 4.

If it is what is called a Crystal Oscillator which is in most timepieces in the form of a Piezo Crystal then it would have to be extremely small. The Crystal Oscillator works on the basis that when presented with an electric current, the crystal distorts and then springs back to its original shape. This sends a timed pulse!

Different Oscillators are separated by their operating frequency and some are more of a full circuit rather than one stand-alone device. They can go from very low hertz to several gigahertz.

Unlike an inverter, the AC signal is pulsed and not constant hence the name, and it is a frequency reliant device.

Standalone Crystal Oscillators would require an Amplifier like a Transistor and a Signal.

Oscillator Wave Forms:

Oscillators are in time devices like watches and timers. There are several types of Oscillator so picking the correct type is essential.

Below are three Wave Patterns that various Oscillators may produce, there are others:

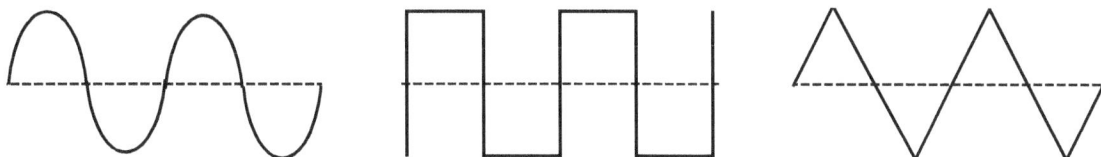

Categories of Oscillator:

There are two categories of Oscillator: 1) Sinusoidal and 2) Non-Sinusoidal, so let us have a look at each one and see what the difference is:

Sinusoidal:

This category produces a signal that is set at a particular constant level and sine wave frequency.

Non-Sinusoidal:

These are different waveforms than Sinusoidal, more like the shapes in the previous diagram of square, triangular etc. Sometimes manufacturers refer to these as Relaxation Oscillators.

Feedback Amplifier:

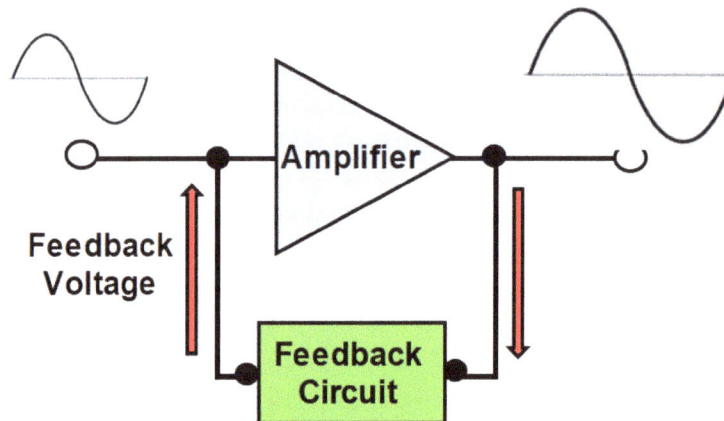

This probably is the most common form of Oscillator. This is where some of the output signal is re-diverted back into the Amplifier input. So you can see by the diagram above how some of the output voltage is fed back to be re-amplified.

We are looking at a feedback voltage combining with the input voltage via a Feedback Circuit Filter to give a much larger output voltage from the Amplifier which, in the end, will sustain the Oscillations.

Types of Oscillator:

As mentioned above it is imperative that you select the correct Oscillator for the task you require it to do! Let us have a look at many types below:

1) MEMS Oscillator

2) Silicon Oscillator

3) Quartz Crystal Oscillator

4) Armstrong Oscillator

5) Hartley Oscillator

6) Phase Shift Oscillator

7) Colpitts Oscillator

8) Cross Coupled Oscillator

9) Dynatron Oscillator

10) Meissner Oscillator

11) Optoelectronic Oscillator

12) Wien Bridge Oscillator

13) Robinson Oscillator

14) Tri-Tet Oscillator

15) RC Phase Shift Oscillator

16) Voltage Controlled Oscillator

17) Clapp Oscillator

18) Gunn Oscillator

19) Tuned Collector Oscillator

20) Cross Coupled Oscillator

21) Ring Oscillator

22) Pierce Oscillator

23) Pearson-Anson Oscillator

24) Delay Line Oscillator

25) Royer Oscillator

26) Electron Coupled Oscillator

27) Multi Wave Oscillator

28) SAW Oscillator

29) CMOS Oscillator

PCB Oscillator Questions and Answers:

Some of the questions and facts below may refer to Oscillators in general rather than just PCB Oscillators.

Q1 – What does an Oscillator do?

A1 – INVERTS DC to AC in short bursts.

Q2 – Does an Oscillator produce a Sine Wave?

A2 – YES but they can produce Square, Saw Tooth or Triangular Wave Forms!

Q3 – Does an Oscillator provide Positive or Negative Feedback?

A3 – POSITIVE.

Q4 – Does an Oscillator produce high or low frequency?

A4 – USUALLY VERY LOW.

Q5 – On what common device would I find an Oscillator?

A5 – RADIO RECEIVERS have special types as an example.

Q6 – Can Oscillators use Negative Feedback?

A6 – YES in rare cases.

Q7 – Does an Oscillator give an output without any input?

A7 – YES MAINLY but there may be a very small pulsed input.

Q8 – Where is a square wave Oscillator used?

A8 – DIGITAL SIGNAL PROCESSING.

Q9 – Can Integrated Circuits contain Oscillators?

A9 – THEY CERTAINLY CAN!

Q10 – Sometimes an Oscillator is termed 'RC Oscillator'. What does this mean?

A10 – There are two meanings here: RC = REFERANCE CLOCK – Most clocks and wrist watches will have Oscillators in them. RC can also mean RESISTOR AND CAPACITOR in a phase shift Oscillator.

Q11 – Sometimes an Oscillator is termed **'LC Oscillator'.** What does this mean?

A11 – LC = INDUCTOR AND CAPACITOR.

Q12 – Is there more than one type of Oscillator?

A12 – YES – There are two types:

> **A – Sinusoidal Oscillators:** These are common and produce a constant amplitude and frequency sinusoidal wave. Can be LC or RC Oscillators. Sometimes referred to as **Harmonic Oscillators.**

> **B – Non-Sinusoidal Oscillators:** These are the ones that produce Square & Triangular Waves. These are sometimes called **Relaxation Oscillators.**

Soldering Section:

If you work at all with electronics, be it building printed circuits or fault finding on them, then one speciality you will require is how to solder correctly. Looking from the outside soldering looks easy, but I can tell you that to do it properly is a fine art.

First of all you would need to know which type of soldering iron, and there are many, would you require to complete the task? Most soldering irons come with a range of tips from a chisel shape to a very fine point.

Which solder is the best for a printed circuit? Again, there are many, ranging from the stick solder a plumber would use, to solder on a small reel with internal flux.

Usually, these days, Lead free solder is used, but remember always try to solder in a ventilated area and wear the correct PPE Also remember that solder sometimes 'spits' so **Eye Protection etc. is essential.**

So which flux is the best for soldering components onto a printed circuit board? Do you require a separate flux, if so which one? Or will the reel type solder with its internal flux be more suited to your task?

If I am de-soldering i.e. removing components from a printed circuit board, are there any easy ways to remove the excess solder?

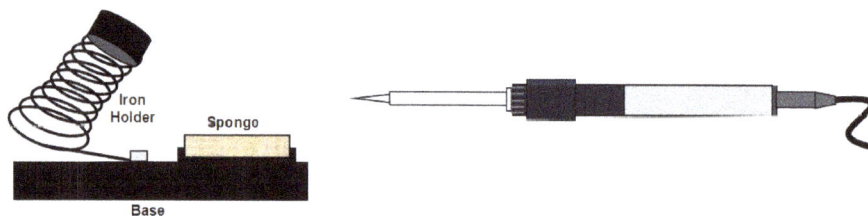

In the diagram above there is a soldering iron holder. Remember this iron is extremely hot and putting it down on a desk or table could be hazardous. I have seen a 240 Volt iron put down haphazardly that melted its own cable!

Looking at the holder, it has a yellow sponge. What is this for, and do I have to soak it in water for any reason? The answer is yes, as the soldering iron tip picks up residue from the soldering such as oxides, pvc off cables etc., it is very quick to wipe the tip of the iron on the damp sponge. Remember to keep the tip fluxed and solder covered!

A pair of tweezers is a handy tool to place very small components into the through holes prior to soldering.

All of the above points and questions will be answered in this section.

Solder for Printed Circuit Boards:

Solder general types:

When soldering components onto a printed circuit board it is extremely important to use the correct solder. One of the important concerns may be the temperature your soldering iron reaches, as some Lead free solders require high temperature soldering irons.

Many people think that solder is solder so the type that plumbers use is ok to be used on a Printed Circuit Board, but I can assure you that is **NOT** the case!

Some solders contain flux within the solder itself (Flux Core Solder) and sometimes the flux has to be added by the person doing the soldering. There are certain fluxes which are not suitable for Printed Circuit Boards, such as those used in plumbing, and we will discuss this later.

Many Lance Type Soldering Irons reach 180°C - 400°C, Small Cordless Soldering Irons reach 200° - 480°C. These temperatures may depend upon the make and cost of the iron, and the type of tip you are using, so check very carefully on purchase.

Solder Type and Metal Content (Lead Based):

Solder is an alloy of metals that **a)** Melt easily and **b)** give mechanical strength to the soldering. The most easily melted solder is Lead based, but there is a health risk here in the fact that using Lead you would be breathing in the Lead fumes. This type of solder may require only 190°C.

The common Lead based solder was Tin/Solder. It could be obtained with different percentages of the metals such as 60% Tin & 40% Lead or 50% Tin and 50% Lead and so on.

As above, as well as the soldering you must look at the mechanical strength that different solders give to components on the Printed Circuit Board. The higher the Tin content of the solder the better the mechanical strength.

I personally still think Lead based solder is the best, but we must remember the safety issues with the Lead fumes. This solder can still be readily obtained for people willing to work with it. All I can say is that if you do use Lead based solder then do it in a well ventilated area and keep your face well away from the track being soldered.

Solder Type and Metal Content (Lead Free):

Another type of solder is Lead Free Solder. Tin is usually one of the metals in the solder alloy for two reasons: **a)** Tin is used because of its low melting point compared with other metals except Lead, and **b)** Tin gives good mechanical strength. Sometimes Silver is used for its low corrosion and excellent conductivity.

Our only problem with Lead free solder, if it is a problem, is that the melting point will be much higher than the Lead based and we could be looking at around 217°C. Any soldering irons, no matter what type, may have to reach this temperature and slightly beyond to be effective.

One of the most common Lead Free solders is 96.5% Tin and 3.5% Silver.

Other Materials & Metals that SOME Solders may contain:

Metals in a Solder Alloy may be: Antimony, Bismuth, Copper, Nickel, Cobalt, Indium, Lead, Silver, Tin, Zinc, Germanium, Rare-Earths, Phosphorus, Aluminium, Arsenic, Gold, Iron & Sulphur.

Now many of these metals may not be suitable to use as far as health and safety goes, or in fact for you to use on a printed circuit board (PCB) as they may cause problems with the track.

Eye Protection:

Always use eye protection when soldering as when the solder melts it has been known to 'spit', especially if has got internal flux. The latter will damage your eyes so please keep safe!

Solder Flux:

General:

The problem with most soldering is Oxides which collect on most metals; for instance if we take Iron, we get Ferrous Oxide which is of course rust. When looking at a printed circuit board the track is made of Copper so Oxides can form here. If we try to solder without flux in the presence of Oxides then the solder will fail without doubt! This may cause what is known as a 'Dry Solder Joint'.

So where does our flux come in? Well on a Printed Circuit Board flux makes the solder run and stick to the copper track more easily. It also removes any coating from the track. Many solders on coils contain flux, so separate flux is not normally necessary, but sometimes advisable. If you use separate flux try and obtain the cream type.

Types of Flux:

Let us have a look at three types of flux and their suitability:

A) **Rosin Flux:** back in the past this flux was made from refined pine sap. This is an Acid based flux when in its liquid form i.e. heated with a soldering iron, and turns inert as it solidifies. This in turn will remove any oxides from the surface being soldered. You might remove any flux residue on completion of soldering, but as it is inert when solid it can be left. Ideal for Printed Circuit Boards.

B) **Water Soluble Organic Flux:** This again is an Acid based flux and is one of the more common, but more aggressive than Rosin Flux. This flux, similar to Rosin, will remove Oxides from the surface. It is advisable to remove any flux residue on completion of soldering. This might involve soap & water, compressed air etc. Can be used on Printed Circuit Boards but may take longer due to having to remove excess flux.

C) **Inorganic Flux/No Clean Flux:** This flux is slightly different and contains a much stronger acid (Hydrochloric), more at home with stronger metals such as Brass & Stainless Steel. You must clean off the residue in this case as it will be detrimental to the solder joint. Not the best for Printed Circuit Boards as any flux residue could be detrimental to the solder and track! Note: Plumbing Solder is far too vicious and abrasive for Printed Circuit Boards.

Soldering without Flux:

It would be very unwise to try and solder without any form of flux at all because all that would happen is that the oxides on the track would prevent the solder from adhering and ultimately you could end up with a condition similar to what is called a **'dry joint'** where there is no continuity between the track and the solder. **These type of faults are extremely difficult to locate.**

It is possible to use alternatives to the flux but the effectiveness may not be quite as good. One such alternative which springs to mind is Vaseline (Petroleum Jelly) although I have not actually used it, I know people who have. I have also known people use Citric Acid to remove the Oxides but the surface may have to be prepared first. I must stress that I have always used flux.

Solder with Internal Flux:

Coils of solder with internal flux can be obtained for soldering on Printed Circuit Boards so that a separate flux is not required unless you wish to use one, in which case the paste flux might be preferable.

The one to buy for Printed Circuit Boards is a coil of Lead Free Solder, probably with a Rosin core, melting point 185°C to 277°C depending upon the manufacturer and size. Usually these coils, depending upon the manufacturer, are something in the region of: Tin 97%, Silver 0.3% & Copper 0.7%. Flux 2%. **The one common denominator is Rosin Flux.**

Soldering:

Clean the soldering iron tip on the damp sponge, then solder and flux the tip first for best results.

Soldering Iron Types:

General:

Which Soldering Irons are the best to use with electronics? The main thing that would be required is a fine tip. On my Electronics Courses, the lance type, which comes with a base and a spring holder to house the iron when getting hot or not in use, was used.

Lance type Soldering Iron:

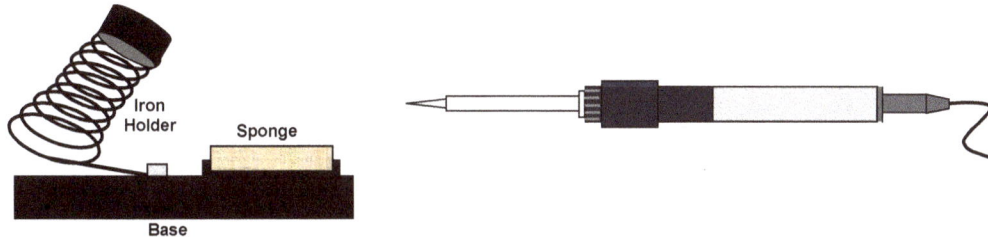

240 volt 'Lance' type Soldering Irons (above) are extremely handy if a lot of soldering is being done. They can be obtained with a number of tips including a very fine one especially for electronic & PCB work. The sponge is dampened on the base to allow you to clean the tip regularly.

Usually, as at the top, they are bought together with a base which includes a spring holder for when the Soldering Iron is heating up or not in use but still hot. They can be obtained with a selection of features, for instance displaying the variable temperature of the tip, and push button switches on the shaft to switch the tip on and off etc. Up to 400°C depending upon the make, cost & type of tip.

Gun Type Soldering Iron:

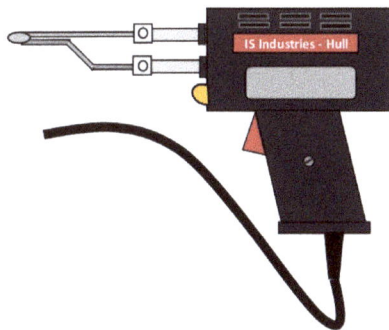

Gun type Soldering irons shaped like the diagram to the left are very handy indeed. When the red trigger is operated the tip will achieve maximum heat in seconds, cooling down when you let go. We used to nickname them a **'quickie'** Soldering Iron because of the speed they heated up. Up to 482°C depending upon make, cost & type etc.

Many of them have an LED light on the front which is shown in yellow on the diagram and this lights up the area where the tip is working. These are usually 240 volts so require an available power supply. Available with very fine tips for electronic and PCB work.

Rechargeable Soldering Iron:

The last Soldering Iron that I would like to mention is the Cordless Soldering Iron similar to the diagram above. One of these would be a very handy addition to your electronic/PCB kit. The time taken to achieve full heat would be around 30 seconds. Takes around one and a half hours to charge and provides around one hour of work time.

This cordless Soldering Iron charges its Lithium Battery via a USB lead and they come with a very fine tip. Similar to the Gun type there is an LED light on the front which shines on the area of the tip. They only take seconds to heat up. Up to 480°C depending upon the make, cost etc.

Solder Removal:

General:

It is all very well talking about how to solder components onto a printed circuit board, but what if we want to remove them from the board – how is that done? Below I have shown three methods of how to remove the solder from 'through hole' components.

Solder Sucker:

A Solder Sucker, sometimes called a De-solder Pump, is a device about 8" long and can be obtained in many different designs. I have just based my diagram on my own personal model.

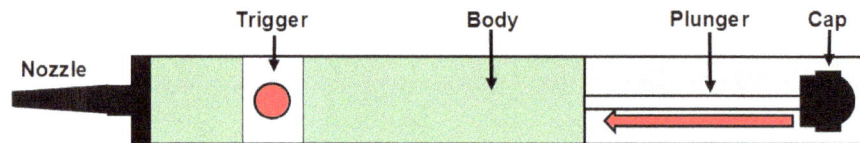

Above is a diagram of a Solder Sucker. The idea is that you place your thumb on the cap and push the plunger into the body of the device as per the red arrow until it latches. By pushing the red button it will de-latch causing a good suction at the nozzle tip.

So if we want to remove a component turn the Printed Circuit Board upside down, turning the PCB over and putting the solder at the top. Take the soldering iron and melt the solder of the through-hole component that you want to remove and at the same time put the tip of the nozzle almost touching the molten solder and press the red trigger release button.

The above action will suck the molten solder into the body of the sucker leaving the wire of the 'through hole' component free of solder. Every now and again the solder will have to be removed from inside of the body and this is done by unscrewing the nozzle. Again, a different size or spare nozzles can be fitted.

De-soldering Bulb:

Very similar to the solder sucker above, except in this case a bulb is squeezed to cause suction at the nozzle.

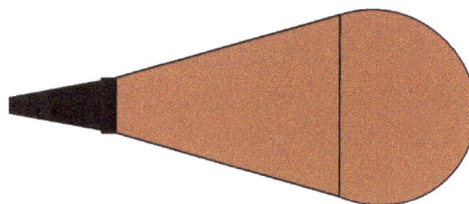

The above method is the cheapest way to do things but I must say is very effective. The nozzle is removed from the end of the bulb to remove the bits of solder.

Solder Wick:

Another way of removing solder from a Printed Circuit Board is with what is called **'Solder Wick'**. This is a very fine copper braid that is placed between the soldering iron and the solder. The braid is fluxed on one side.

So if we want to remove a component, turn the Printed Circuit Board upside down, putting the solder at the top. Take the soldering iron and melt the solder of the through-hole component that you want to remove then place the braid on the solder and replace the soldering iron tip back onto the braid.

As the iron heats up the braid, the solder will rise up the braid very similar to capillary action. The braid is on a coil so cut off the part you use and it is ready for the next task.

Soldering Questions & Answers:

Q1 – Do I need to use flux when soldering printed circuits?

A1 – **YES.** To remove oxidation on the track otherwise it would be very difficult for the solder to adhere! Some coil solders have internal flux.

Q2 – Can a soldering iron not get hot enough?

A2 – **YES** – but that would be a rarity rather than common. The iron would be ideal around 200°C+ to 400°C+ usually depending on the tip.

Q3 – What is the yellow sponge for on the base?

A3 – **CLEANING** the tip of the iron. The sponge should be moist.

Q4 – Which Flux is the best for soldering Printed Circuit Boards?

A4 – **ROSIN FLUX.**

Q5 – Which solder is the best?

A5 – **LEAD FREE** with internal Rosin Flux is the safest Health & Safety wise.

Q6 – Will plumbers' solder be ok for Printed Circuit Boards?

A6 – **CERTAINLY NOT.** Much too vicious & abrasive for the track.

Q7 – How long will a Cordless Soldering Iron charge last?

A7 – **AROUND 1 HOUR.**

Q8 – Can Cordless Soldering Irons be used on any solder?

A8 – **NO** usually used on coil solder which is less than 1.5mm.

Q9 – Can I remove solder from a PCB using Acetone?

A9 – **CERTAINLY NOT.** Although doing this may in fact work, and some papers actually advise, using chemicals on the material of a PCB is not advised as they can be detrimental to the board make up. If you are intent on using chemicals on a PCB to clean off excess debris or solder, seek advice first!

Q10 – Is there an alternative to a solder sucker for removing solder?

A10 – **YES.** Solder Wick which is a very thin braid placed between the solder and the soldering iron. This solder wick comes on coils and soaks up the solder.

Q11 – Are all coil solders the same?

A11 – **NO.** Some contain just Tin and Copper, some are Tin, Copper and Nickel, whilst others are just Tin and Silver etc. They all should contain Rosin Flux.

Q12 – Do I just drop the molten solder onto the wire/track?

A12 – **CERTAINLY NOT.** The soldering iron must be full heat and in contact with the track and wire, surface mounted components may require a heat sink! When the wire and track are hot apply the solder/flux. The soldering iron should be the first item to touch the track and the last to be removed.

Building a Circuit Section:

For anyone who is not familiar with soldering I have drawn a circuit that I made some time ago which is a small inverter. Will quite easily be housed on Vero-board.

Not too many components required and the Variable Resistors allow you to adjust the output voltage when the power is applied.

When you get a bit more experienced it is always a good challenge to design your own circuit, start small and evolve your circuit. Just start with an LED, a Battery and several Resistances on Vero-board and build the circuit up from there.

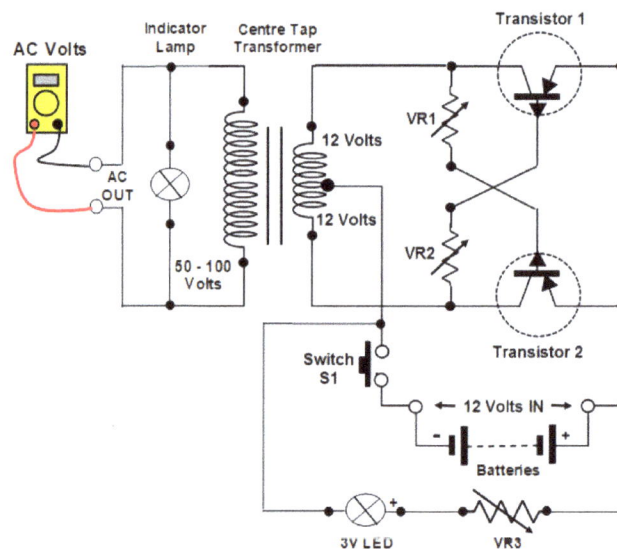

Starter Equipment Recommended:

1) Multi-meter – Preferably with diode testing facility.

2) Soldering Iron – With fine tip.

3) Solder – Lead free coil with internal Rosin Flux.

4) Solder Remover – Inexpensive Bulb type.

5) Magnifying Glass – Preferably with stand and holder.

6) Breadboard – Not necessarily an expensive one with integral batteries etc.

7) LED Kit – Can be obtained cheaply 3 – 6 Volt will be suffice.

8) Resistor Kit – Can be obtained cheaply with a good range of Resistances.

9) Tweezers – Some components are very small and fiddly.

10) Vero Board – With around 10 tracks.

11) 6 – 9 Volt Battery (something like a PP3).

The above equipment will start you off, as well as a selection of tools. You can become familiar with soldering components onto the Vero Board and removing them after testing them on the Breadboard with the Multi-meter.

Basic Inverter Circuit:

If you wanted to have a go at building an electronic circuit, a simple one might be the inverter below. The candidates on our course had to design a circuit schematic on the final day of the course (with a little help).

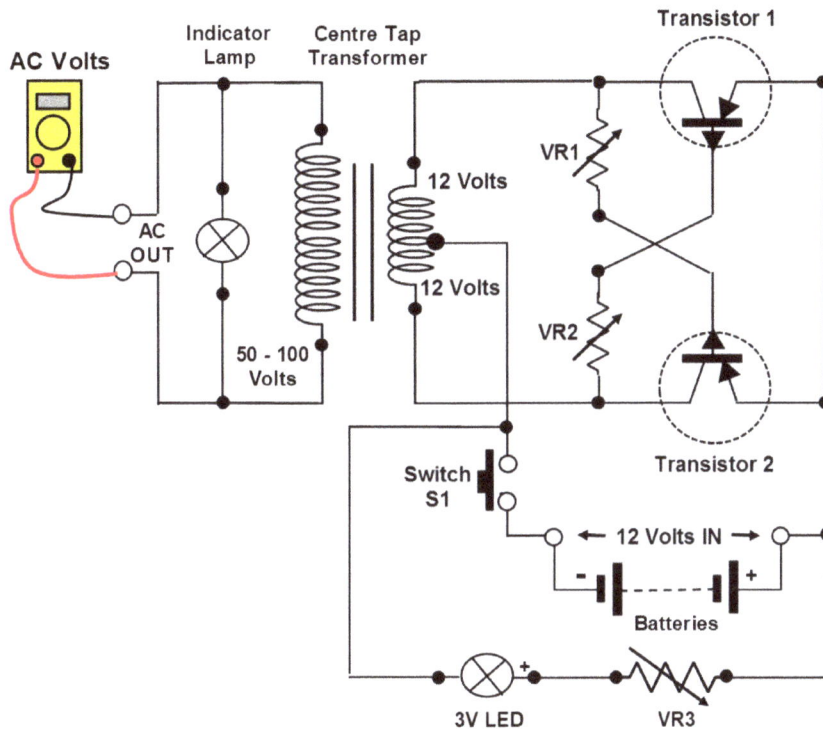

I will list the components that you would require:

1) 2 x PNP Power Transistors.

2) 2 x Variable Resistors (V1 & V2) 60Ω - 80Ω

3) 1 x 12 Volt Battery.

4) 1 x Centre Tap Transformer. 12 Volt to whatever output you like.

5) 1 x 3 Volt LED

6) 1 x Variable Resistor/Static Resistor (VR3) (around 20KΩ)

7) 1 x Indicator Lamp (output) Voltage depends upon your output voltage.

8) 1 x Pushbutton Switch.

9) Vero-board of enough size for the components (10 Tracks wide).

10) 1 x Multi-meter.

When setting the Variable Resistors (V1 & V2) try setting them around 70Ω to start with. You can experiment with different components; for instance I have used PNP Transistors, but try designing a circuit with NPN Transistors and see what you come up with. Once you have got the Variable Resistances set to an ideal value you can use Static Carbon Resistors.

I have wired in a Multi-meter on the diagram. Ensure that it is set to AC volts to check the output so that adjustment on the variable resistors can be checked. I have laid the circuit out on Vero-board next.

Inverter Building on Vero-board:

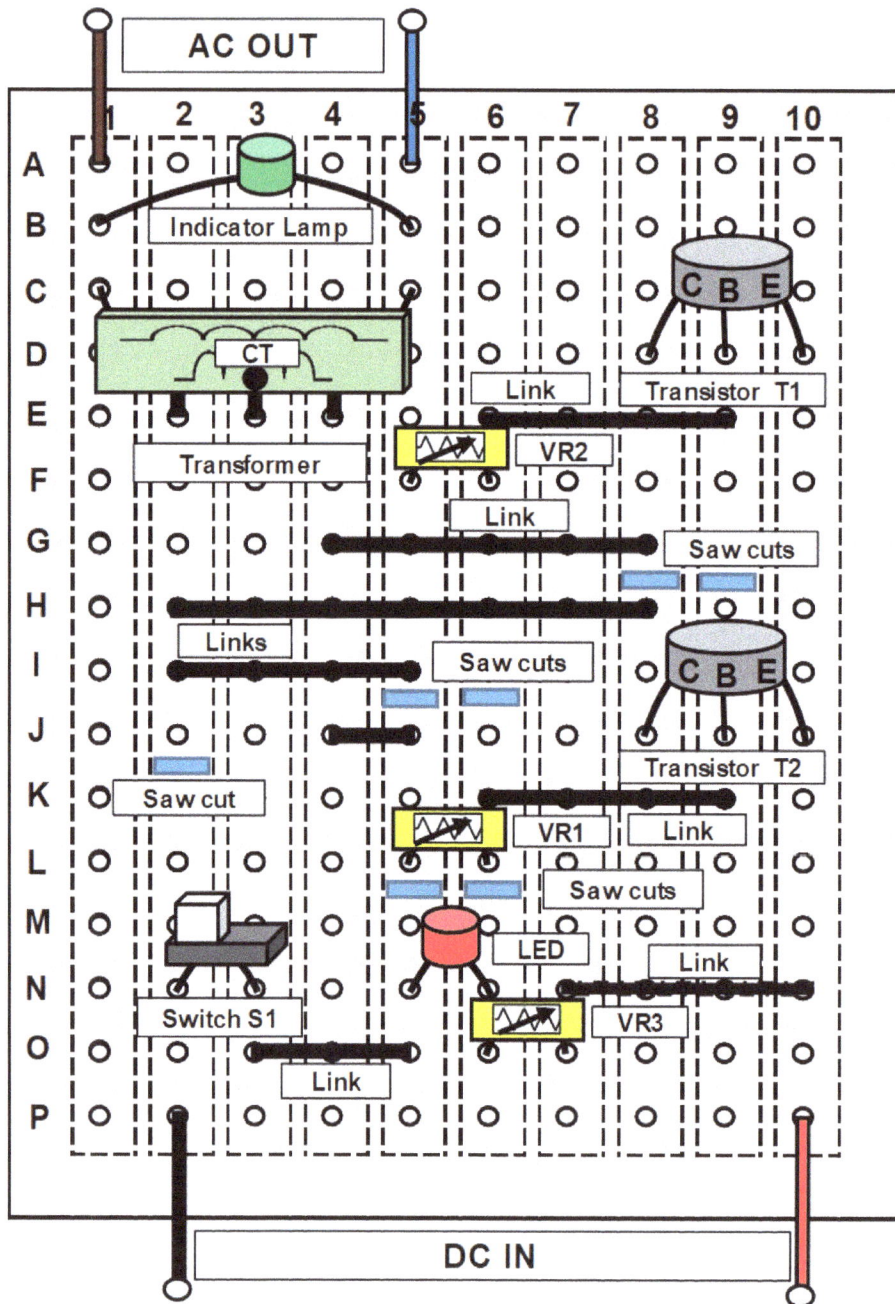

1) As a project to practice soldering I have laid the previous Inverter Schematic Diagram out on Vero-board above. Remember when you turn the Vero-board over to solder, components will be on the other side of the board, i.e. where they were on the right on the top of the board, such as the Transistors, they will be on the left when you turn the board over.

2) Sometimes, for the first attempt, it helps if you take a very fine black permanent marker and mark the Vero-board with numbers and letters as above, similar to a 'breadboard'.

3) It is a simple project that after the DC is applied, the Variable Resistors will need adjusting until the AC output is stable at the required voltage.

4) You will see that the blue shapes are saw cuts in the track to separate the components. The thick black lines are links from a – b, not joining the track in between.

Index:

www.ingramcontent.com/pod-product-compliance
Lightning Source LLC
Chambersburg PA
CBHW041620220326
41597CB00035BA/6181